Graphene and Its Derivatives - Synthesis and Applications

*Edited by Ishaq Ahmad
and Fabian I. Ezema*

Published in London, United Kingdom

IntechOpen

Supporting open minds since 2005

Graphene and Its Derivatives - Synthesis and Applications
http://dx.doi.org/10.5772/intechopen.73354
Edited by Ishaq Ahmad and Fabian I. Ezema

Contributors
Ning Wang, Haixu Wang, Guang Yang, Rong Sun, Ching-Ping Wong, He Tian, Yi Yang, Lu-Qi Tao, Fan Wu, Guang-Yang Gou, Tian-Ling Ren, Humaira Seema, Randhir Bhoria, Ishaq Ahmad

Notice
Statements and opinions expressed in the chapters are these of the individual contributors and not necessarily those of the editors or publisher. No responsibility is accepted for the accuracy of information contained in the published chapters. The publisher assumes no responsibility for any damage or injury to persons or property arising out of the use of any materials, instructions, methods or ideas contained in the book.

First published in London, United Kingdom, 2019 by IntechOpen
IntechOpen is the global imprint of INTECHOPEN LIMITED, registered in England and Wales, registration number: 11086078, 7th floor, 10 Lower Thames Street, London, EC3R 6AF, United Kingdom
Printed in Croatia

British Library Cataloguing-in-Publication Data
A catalogue record for this book is available from the British Library

Additional hard and PDF copies can be obtained from orders@intechopen.com

Graphene and Its Derivatives - Synthesis and Applications
Edited by Ishaq Ahmad and Fabian I. Ezema
p. cm.
Print ISBN 978-1-83962-881-8
Online ISBN 978-1-83962-882-5
eBook (PDF) ISBN 978-1-83962-883-2

We are IntechOpen,
the world's leading publisher of
Open Access books
Built by scientists, for scientists

4,400+
Open access books available

117,000+
International authors and editors

130M+
Downloads

Our authors are among the

151
Countries delivered to

Top 1%
most cited scientists

12.2%
Contributors from top 500 universities

Interested in publishing with us?
Contact book.department@intechopen.com

Numbers displayed above are based on latest data collected.
For more information visit www.intechopen.com

Meet the editors

Ishaq Ahmad joined the National Centre for Physics in 1999 and currently holds the positions of Director of the Experimental Physics Department, National Centre for Physics (Islamabad) and Co-Director of the NPU-NCP Joint International Research Centre on Advanced Nanomaterials and Defects Engineering, Northwestern Polytechnical University, Xian, China. He has co-authored over 120 publications in SCI journals and supervised several Master, PhD, and Postdoctoral students. He has authored a number of books and book chapters. His research interests are focused on ion implantation, radiation-induced modification of materials/nanomaterials, synthesis of nanomaterials/thin films, and ion beam analysis of materials. He is an editorial board member and reviewer of several well-known journals. He was a TWAS-UNESCO-iThemba LABS Research Associate from 2014 to 2016. He obtained his PhD at the Graduate University of Chinese Academy of Sciences, China.

Fabian Ezema is a Professor at the University of Nigeria, Nsukka. He obtained a PhD in Physics and Astronomy from the University of Nigeria (UNN), Nsukka in 2000. His research focuses on several areas of Materials Science: synthesis and characterizations of particles and thin film materials on energy and sensing applications. He was a CV Raman Fellow at the Shivaji University, Kolhapur, India in 2011 and an MIF Fellow at the Tokyo University of Science, Japan in 2013. He is currently a visiting professor to NRS-EMT (MATECSS UNESCO Chair) in Varennes, Quebec, Canada and iThemba Labs, South Africa. He is also a Fellow of UNESCO-UNISA South Africa and the *Chair in Nanosciences and Nanotechnology* (U2ACN2). He has been awarded the UNISA visiting researcher for 12 months to work on perovskite solar cells. He is the team leader of the Nano Research Group at UNN. He is an Acting Director of the International Office of the University of Nigeria. He is the Pioneer Dean of the Faculty of Natural and Applied Sciences, Coal City University, Enugu (on sabbatical leave). With international collaborators from Canada, USA, Japan, South Africa, Singapore, Malaysia, India, Pakistan, etc, Fabian has published over 140 high impact papers in various national and international journals and has presented over 30 talks at various conferences, workshops, and seminars.

Contents

Preface

Presently, graphene is widely researched worldwide because of its unique properties, which have led to a wide range of applications. This book provides a brief overview of recent developments in the synthesis methods of graphene and its derivatives as well as its applications.

The review in the first part of the book covers the recent synthesis methods of graphene and its modified derivatives that include graphene oxide (GO), graphene (rGO), graphene-metal nanoparticle composites, graphene-polymer hybrids, and graphene/organic structures for a variety of applications such as catalysts, energy storage/conversion, anti-microbial agents, and as a water decontaminant. Also presented in this section is a unique and advanced technique that is the liquid phase exfoliation method for the synthesis and concentration enhancement of graphene with the addition of certain additives and salts. This method is suitable for the enhancement of the concentration of graphene. This process can be easily scaled up for better performance and efficiency to be used for the fabrication of modern electronic devices.

The final section of this book addresses the most important applications of graphene and its derivatives. Discussed in detail in this book are photocatalytic applications, electronic applications, and the latest graphene-based heterogeneous electrodes for energy storage. In addition, sound devices based on graphene are also presented in this book.

The authors are very thankful and want to acknowledge all those who contributed to this book: Prof. Dr. Seema Humaira, Prof. Dr. Randhir Singh, Prof. Dr. Ning Wang, Prof. Dr. Haixu Wang, Prof. Dr. Guang Yang, Prof. Dr. Rong Sun, Prof. Dr. Ching-Ping Wong, Prof. Dr. He Tian, Prof. Dr. Guang-yang Gou, Prof. Dr. Fan Wu, Prof. Dr. Lu-Qi Tao, Prof. Dr. Yi Yang, Prof. Dr. Tian-Ling Ren, Dr. R . M. Obodo, and Ms. Manuela Gabric, Author Service Manager.

Ishaq Ahmad
NPU-NCP Joint International Research Center on Advanced Nanomaterials and
Defects Engineering,
National Centre for Physics,
Islamabad, Pakistan

Fabian I. Ezema
Department of Physics and Astronomy,
University of Nigeria,
Nsukka, Nigeria

Coal City University,
Enugu State, Nigeria

Synthesis of Graphene and Its Derivatives

Chapter 1

Introductory Chapter: Graphene and Its Applications

Raphael Mmaduka Obodo, Ishaq Ahmad
and Fabian Ifeanyichukwu Ezema

1. Introduction

Presently, graphene is widely researched worldwide because of its unique properties such as zero bandgap, remarkable electron mobility at room temperature, high thermal conductivity and stiffness, large surface area, impermeability to gases, etc. Graphene charge carrier exhibits core mobility, is massless, and moves a few micrometers distance maintaining its structure at room temperature. Recently, graphene-based materials have gained intense awareness on energy storage systems, electronics, chemical sensors, optoelectronics, nanocomposites, and health such as osteogenic. Graphene is among the allotropes of carbon; its carbon atoms are arranged in a single layer. These carbon atoms are organized in a honeycomb lattice with a two-dimensional arrangement. The carbon–carbon bond distance in a single graphene sheet approximates 0.142 nm [1]. One of the unique and major properties of graphene is that increased researcher's interest is its constituent's electrons that seem to be massless relativistic particles, hence, anomalous quantum Hall effect and the absence of localization [2, 3]. Graphene has been used in many applications, which include energy storage devices like supercapacitors and lithium-ion batteries [4], gas detection [5], and conducting electrodes [6]. Recently, the rate at which graphene awareness is rising is highly remarkable and suggests that it is a good route of the scientists' search for new materials for advancement in science, engineering, health, and composite industries. This brief introduction of graphene narrates its brief history, synthesis method, derivatives, and applications. Addition of graphene in a composite inhibits the fabrications of active material in a nanosize, enhances non-faradaic capacitive behavior, increases conductivity, and prevents disintegration. Graphene also induces a physical barrier in between the electrolyte and active material, hence increasing cycling stability, specific capacitance, and rate capability.

2. Synthesis of graphene

Graphene synthesis means any process of fabricating or extracting graphene from graphite. The method to be chosen is governed by the desired size, quantity, and purity. Synthesis technique contributes to the structure and properties of graphene produced. There are variations of graphene layers from different techniques such as a single layer, double layer, or multiple layers, and they have different applications in various fields of science and technology like energy storage devices, biotechnology, memory, electronics, sensors, etc. Researchers employ different techniques especially when a large quantity is required. Subsequently, we will discuss various synthesis techniques, applications, its status now, progress so far, and future prospects.

In the synthesis of graphene-based materials, ball milling and hydrothermal methods show to be cheaper, the electrospinning method exhibits the benefits in the nanowire composite assembly, and the microwave-assisted method is easier and superfast in fabrication. We also explained methods of graphene synthesis while its derivatives are discussed in the second chapter of this book. The third chapter explained the new technique such as liquid phase exfoliation method for the synthesis and concentration enhancement of graphene which is suitable for the fabrication of the highly efficient modern electronic devices (**Figure 1**).

2.1 Cleavage and exfoliation technique

This method is divided into two: (1) mechanical exfoliation and (2) chemical exfoliation. Mechanical exfoliation is the distortion of weak van der Waals force holding carbon–carbon atom together. The chemical method is the production of colloidal suspension which produces graphene from graphite compounds. Graphite is several densely packed layers of graphene sheets, hence, fixed together by weak van der Waals force. High-purity graphene sheets can be produced from graphite sheet by breaking the bonds that held them together. Therefore, exfoliation and cleavage are the use of mechanical or chemical energy to break down these weak bonds and separate distinctive graphene sheets. Viculis et al. [4] were the first to apply this principle by using potassium metal to separate pure graphite sheet and then exfoliate them using ethanol to form a dispersion of graphene sheets.

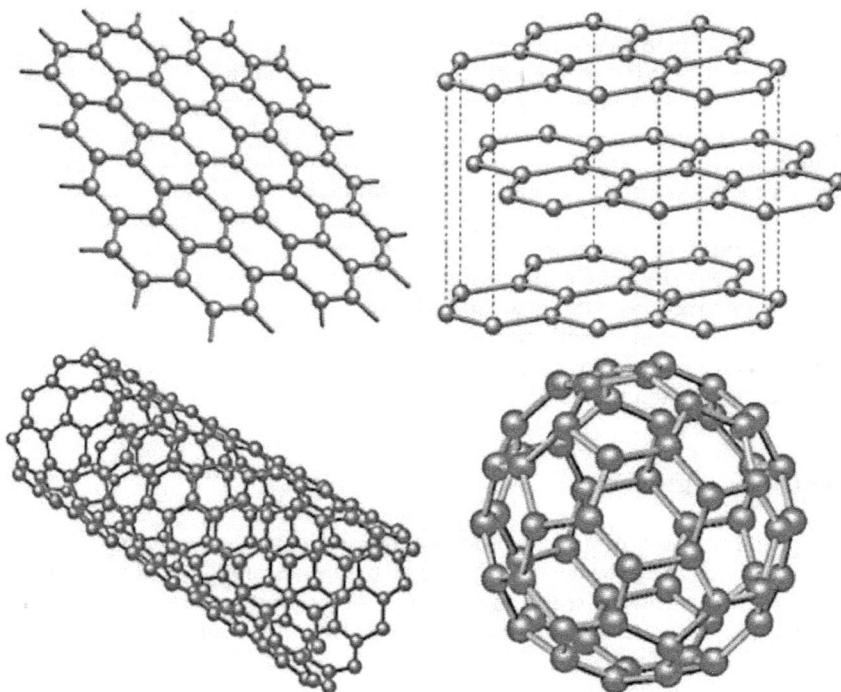

Figure 1.
Structure of graphene sheet, stacked graphene, wrapped graphene, and rolled graphene. Reproduced from Ref. [7].

2.2 Chemical vapor deposition (CVD) methods

Chemical vapor techniques use steam phase exfoliation. This method chemically extracts graphene sheets from graphite without passing through exfoliation stage. Horiuchi et al. [9] were the first people to produce graphene sheets using this method. They engaged the method to fabricate carbon nanofilms (CNF) using regular graphite sheets.

There are many types of CVD, depending on the precursors available, the structure needed, quality of material, and dimension, and there are many applicable CVD processes such as thermal, plasma-enhanced (PECVD), cold wall, reactive, hot wall [9], etc. Graphene thin films are formed on copper or nickel mostly by a chemical vapor deposition method.

2.3 Pyrolysis of graphene

The pyrolysis uses solvothermal technique to synthesize graphene from graphite by a bottom-up approach. Sodium ethoxide and ethanol were mixed in a molar mass ratio of 1:1 in a closed vessel with intense heat treatment and sonication; this process detaches graphene from graphite [10].

2.4 Other techniques

2.4.1 Unzipping CNTs

One of the most recent techniques of fabricating graphene is a type of synthesis that uses multiwall carbon nanotubes (MWNT) as initial material. This method is commonly known as **"CNTs' un-zipping."** MWNTs can be unzipped longitudinally using lithium and ammonia intercalation, followed by intense acid and heat treatment, which induces exfoliation immediately [11].

2.4.2 Thermal decomposition of ruthenium crystal

Graphene single layers can be grown on single crystal ruthenium (Ru 0001) surface at ultra-high vacuum (4.0×10^{-11} Torr) [11]. It was discovered that graphene could form on the crystal surface. This can be achieved by heat breakdown of ethylene (pre-adsorbed on the crystal surface at room temperature) at 1000 K or by controlled segregation of carbon from the bulk of the substrate [12].

2.4.3 Thermal decomposition of SiC

Thermal disintegration of silicon on the surface plane of a single crystal of 6H-SiC to produce graphene recently gained researchers' awareness. It takes less time to achieve and become popular techniques of graphene growth recently [13].

3. Graphene oxide

Graphene oxide (GO) is a product of graphene obtained by oxidizing graphene. It has a single monomolecular layer containing oxygen functionalities such as carboxyl, carbonyl, epoxide, or hydroxyl groups [14]. These added functionalities expand the separation between the layers and make the material hydrophilic (meaning that they can be dispersed in water). Layers of graphene stacked on top of each other form graphite, with an interplanar spacing of 0.335 nm. The separate layers of graphene in graphite are held together by van der Waals forces. GO are synthesized mostly based on widely reported Hummers method in which graphite is oxidized by a solution of potassium permanganate in hydrogen tetraoxosulfate (IV) acid [15].

The diagram in **Figure 2** illustrates the processes and stages involved in moving from graphite to graphene, graphene to graphene oxide, and graphene oxide to reduced graphene oxide [14, 16]. Many scientists are confused about the difference between carbon derivatives (**Figure 3**).

Graphene oxide is dispersible in water and other organic solvents like ethanol, 1-propanol, acetone, methanol, ethylene glycol, pyridine, etc. as well as in different matrixes. This property of GO was due to the presence of the oxygen functionalities.

Modification of Graphite powders

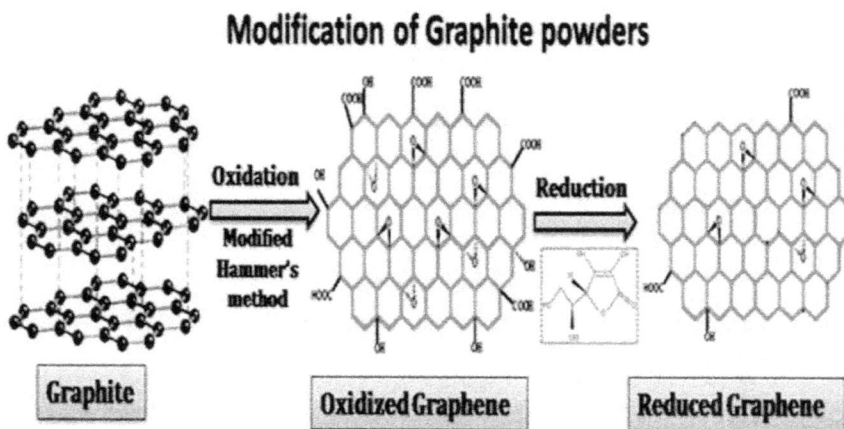

Figure 2.
Stages of synthesis of GO and rGO. Reproduced from Ref. [8].

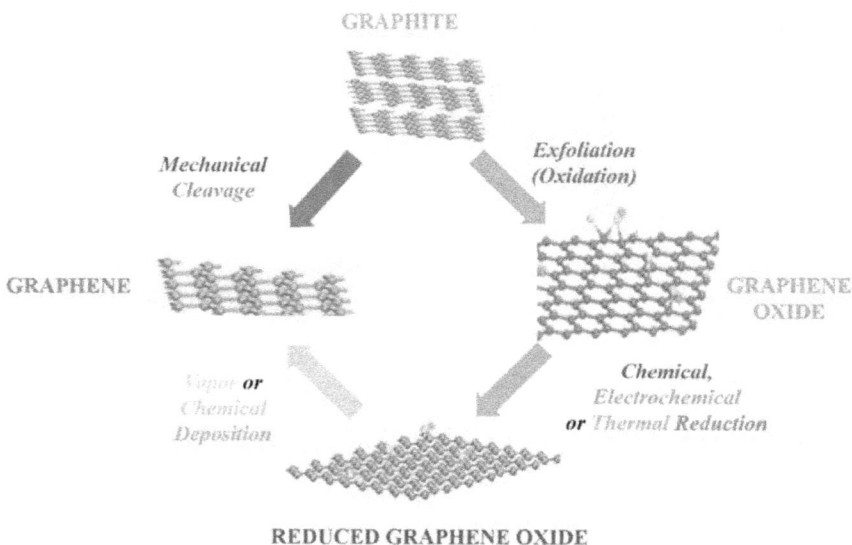

Figure 3.
Cycle synthesis of graphene/GO/rGO. Reproduced from Ref. [33].

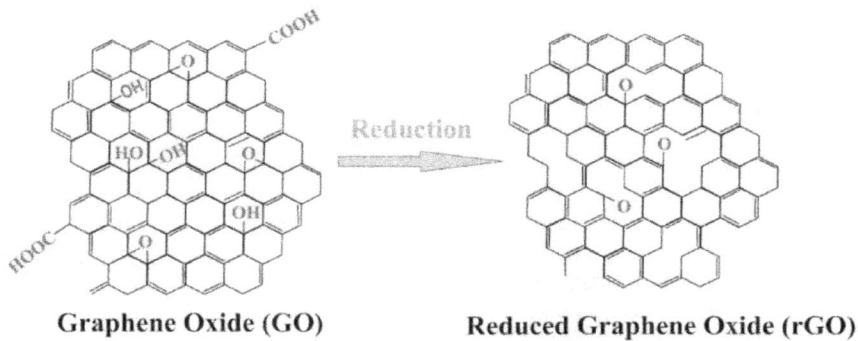

Graphene Oxide (GO)   **Reduced Graphene Oxide (rGO)**

Figure 4.
Diagram of reduced graphene oxide (rGO). Reproduced from Ref. [34].

3.1 Reduced graphene oxide (rGO)

Reduced graphene oxide (rGO) is a graphene oxide (GO) in which its oxygen content is reduced either by thermal, chemical, or any other methods. Graphene oxide is reduced to improve the honeycomb hexagonal lattice distorted during oxidation from graphene to graphene oxide and also enhance its electrical conductivity [14, 34]. It is also observed that once most of the oxygen groups are removed, the reduced graphene oxide obtained becomes indispersible in a solvent due to its tendency to create aggregates (**Figure 4**).

4. Applications of graphene/GO/rGO

Graphite and its derivate recently gained science and engineering awareness due to its numerous applications. The discovery of graphene is rightly regarded as a milestone in the world of material science; as can be seen in the worldwide attention, the material has received in the fields of electronics, photonics, capacitors/supercapacitors, biosensing, etc. They are used in numerous applications as illustrated below. In this book, applications of graphene and its derivatives are discussed in detail. These applications include photocatalysis, electronics, gas sensing, graphene-based heterogeneous electrodes for energy storage devices, etc. In addition, sound devices based on graphene is also explained in this book.

4.1 Electronics

GO are used in electronic fabrications as initial materials. Electronic devices such as graphene effect transistors (GFETs) and field effect transistors (FETs) are graphene-based [17]. Reduced graphene oxides (rGO) are used as chemical sensors [18]. Functionalized graphene oxide in conjunction with glucose oxidase deposited on electrode material is used as an electrochemical glucose sensor [19]. They are widely used in the manufacturing of electronic devices like light-emitting diodes (LEDs) and solar cells. Reduced graphene oxide dispersed in a solvent can be used in the production of the transparent electrode, which is an alternative transparent electrode like FTO and ITO [20].

4.2 Energy storage

Reduced graphene oxide nanocomposites have a high surface area and good conductivity, which suited them for use in supercapacitors and lithium-ion batteries with good energy storage capacity. GO-based supercapacitors and lithium-ion batteries possess high-energy storage capacity, long life span, and good cycle stability.

4.3 Water purification

As far, back as the 1960s [21], scientists have started studying graphite oxide usage in desalination of water. In 2011, some group of researchers employed the principle of reverse osmosis using GO to achieve the same goal [22]. It was discovered that graphite allows water to pass through but retain some larger ions [23]. Its narrow mono- or bilayer capillaries allow water but restrain heavy ions.

Moreover, in the year 2015, a group of scientists also purified water using graphene tea by removing 95% of heavy metal ions in water solution [24].

It was reported that in 2006, engineers fabricated graphene-based thin film powered by solar energy that possesses the quality of filtering dirty and salty water. These films are non-heavy and can be easily produced on a large scale [24].

4.4 Biomedical applications

Graphite and its derivative like GO are widely used in the biomedical field as a constituent in the drug delivery system. Magnetite stacked with GO and doxorubicin hydrochloride (DXR) drug adsorbed onto the system is used as anticancer treatment by targeting it to a specific site to kill cancer cells.

4.5 Biosensors

Graphene oxide and reduced graphene oxide have been incorporated into many gadgets. These GO-/rGO-based gadgets are fabricated with the quality to identify biologically significant molecules. GO/rGO uses fluorescence resonance energy transfer (FRET) characteristics to work effectively as a biosensor.

4.6 Elemental storage

All elements that form part of GO or rGO functional groups can be effectively stored in their sheets and extracted later for use and are also being explored for their applications in hydrogen storage.

4.7 Plasmonics

Recently, the science of plasmonics discovered that near field infrared optical microscopy [25] and infrared spectroscopy [26] of graphene provide accommodations for plasmonic surface mode [27].

4.8 Lubricant

Scientists recently found out that graphene lubricants perform better than regularly used graphite lubricants. A graphene lubricant applied to a ball and bearing roller or steel ball and steel disc lasted for 6500 cycles, while our usually used graphite lubricants lasted only for 1000 cycles [24].

4.9 Radio wave absorption

A heavenly crammed graphene layer deposited on glass substrates absorbs radio waves of the wavelength range of 125–165 GHz bandwidth by 90% [24]. In our modern houses, graphene serves as roof, door, and window coatings to safeguard houses from radio wave interference [28].

4.10 Nanoantennas

A nanoantenna called graphene-based plasmonic nanoantenna (GPN) operates on a wavelength of millimeter within the radio wavelength range. This nanoantenna is better than our conventional antennas because its operational surface plasmon polaritons wavelength is much smaller compared to the wavelength of electromagnetic waves propagating at the same frequency. Our conventional antenna operational frequencies range from 100 to 1000, which is very huge compared to GPNs [29].

4.11 Sound transducers

Graphene has been predicted as a good candidate for the manufacturing of electrostatic audio microphones and speakers due to their lightweight, which provides moderately good frequency response [30]. In 2015 an A model audio ultrasonic microphone and the speaker was fabricated; it operates at a frequency range of 20–500 kHz [31]. Its performance operation was up to 99% efficiency, good and uniform frequency output throughout the audible range [32].

Author details

Raphael Mmaduka Obodo[1,2], Ishaq Ahmad[2] and Fabian Ifeanyichukwu Ezema[1*]

1 Department of Physics and Astronomy, University of Nigeria, Nsukka, Enugu, Nigeria

2 NPU-NCP Joint International Research Center on Advance Nanomaterials and Defects Engineering, National Center for Physics, Islamabad, Pakistan

*Address all correspondence to: fabian.ezema@unn.edu.ng

IntechOpen

References

[1] Mallard LM, Pimenta MA, Dresselhaus G, Dresselhaus MS. Raman spectroscopy in graphene. Physics Reports. 2009;**473**:51-87

[2] Geim AK, Kim P. Carbon wonderland. Scientific American. 2008;**298**:90

[3] Novoselov KS, Geim AK, Morozov SV, Jiang D, Zhang Y, Dubonos SV, et al. Electric field effect in atomically thin carbon films. Science. 2004;**306**:666

[4] Viculis LM, Mack JJ, Kaner RB. A chemical route to carbon nanoscrolls. Science. 2003;**299**:1361

[5] Berger C, Song Z, Li T, Li X, Ogbazghi AY, Feng R, et al. Ultrathin epitaxial graphite: 2D electron gas properties and a route toward graphene-based nanoelectronics. The Journal of Physical Chemistry. 2004;**108**:19912

[6] Land TA, Michely T, Behm RJ, Hemminger JC, Comsa G. STM investigation of single layer graphite structures produced on Pt(111) by hydrocarbon decomposition. Surface Science. 1992;**264**:261

[7] Castro Neto AH, Guinea F, Peres NMR. Drawing conclusions from graphene. Physics World. 2006;**19**:33

[8] Swain SS, Unnikrishnan L, Mohanty S, Nayak SK. Hybridization of MWCNTs and reduced graphene oxide on random and electrically aligned nanocomposite membrane for selective separation of O2/N2 gas pair. Nayak Journal of Materials Science. 2018;**53**(22):15442-15464

[9] Horiuchi S, Gotou T, Fujiwara M, Asaka T, Yokosawa T, Matsui Y. Single graphene sheet detected in a carbon nanofilm. Applied Physics Letters. 2004;**84**:2403

[10] Obraztsov AN, Zolotukhin AA, Ustinov AO, Volkov AP, Svirko Y, Jefimovs K. DC discharge plasma studies for nanostructured carbon CVD. Diamond and Related Materials. 2003;**12**:917

[11] Bhuyan MSA, Uddin MN, Islam MM, Bipasha FA, Hossain SS. Synthesis of graphene. International Nano Letters. 2016;**6**:65-83

[12] Cano-Márquez AG, Rodríguez-Macías FJ, Campos-Delgado J, Espinosa-González CG, Tristán-López F, Ramíre-González D, et al. Ex-MWNTs: Graphene sheets and ribbons produced by lithium intercalation and exfoliation of carbon nanotubes. Nano Letters. 2009;**9**:1527

[13] Vázquez de Parga AL, Calleja F, Borca B, Passeggi MCG Jr, Hinarejos JJ, Guinea F, et al. Periodically rippled graphene: Growth and spatially resolved electronic structure. Physical Review Letters. 2008;**100**:056807

[14] Mao S, Pu H, Chen J. Graphene oxide and its reduction: Modeling and experimental progress. RSC Advances. 2012;**2**:2643-2662

[15] Hummers WS, Offeman RE. Preparation of graphitic oxide. Journal of the American Chemical Society. 1958;**80**:1339

[16] Raccichini R, Varzi A, Passerini S, Scrosati B. Nature Materials. Boosting the power performance of multilayer graphene as lithium-ion battery anode via unconventional doping with in-situ formed Fe nanoparticles. 2015;**14**:271-279

[17] Wang S, Ang PK, Wang Z, Tang ALL, Thong JTL, Loh KP. High mobility, printable, and solution-processed graphene electronics. Nano Letters. 2010;**10**:92

[18] Chen K, Lu G, Chang J, Mao S, Yu K, Cui S, et al. Hg(II) ion detection using thermally reduced graphene oxide decorated with functionalized gold nanoparticles. Analytical Chemistry. 2012;**84**:4057

[19] Liu Y, Yu D, Zeng C, Miao Z, Dai L. Biocompatible graphene oxide-based glucose biosensors. Langmuir. 2010;**26**:6158

[20] Matyba P, Yamaguchi H, Eda G, Chhowalla M, Edman L, Robinson ND. Graphene and mobile ions: The key to all-plastic, solution-processed light-emitting devices. ACS Nano. 2010;**4**:637

[21] Bober ES. Final Report on Reverse Osmosis Membranes Containing Graphitic Oxide. U.S. Dept. Of the Interior; 1970. 116p

[22] Gao W, Majumder M, Alemany LB, Narayanan TN, Ibarra MA, Pradhan BK, et al. Engineered graphite oxide materials for application in water purification. ACS Applied Materials and Interfaces. 2011;**3**(6):1821-1826

[23] Joshi RK, Carbone P, Wang FC, Kravets VG, Su Y, Grigorieva IV, et al. Precise and ultrafast molecular sieving through graphene oxide membranes. Science. 2014;**343**(6172):752-754

[24] https://en.wikipedia.org/wiki/Graphite_oxide#Water_purification [Accessed: February 22, 2019]

[25] Fei Z, Rodin AS, Andreev GO, Bao W, McLeod AS, Wagner M, et al. Gate-tuning of graphene plasmons revealed by infrared nano-imaging. Nature. 2012;**487**(7405):82-85

[26] Yan H, Low T, Zhu W, Wu Y, Freitag M, Li X, et al. Damping pathways of mid-infrared plasmons in graphene nanostructures. Nature Photonics. 2013;**7**(5):394-399

[27] Low T, Avouris P. Graphene plasmonics for terahertz to mid-infrared applications. ACS Nano. 2014;**8**(2):1086-1101

[28] Wu B, Tuncer HM, Naeem M, Yang B, Cole MT, Milne WI, et al. Experimental demonstration of a transparent graphene millimetre wave absorber with 28% fractional bandwidth at 140 GHz. Scientific Reports. 2014;**4**:4130

[29] https://en.wikipedia.org/wiki/Graphite_oxide#lubricants [Accessed: February 22, 2019]

[30] https://en.wikipedia.org/wiki/Graphite_oxide#micriphones [Accessed: February 22, 2019]

[31] http://research.physics.berkeley.edu/zettl/pdf/471 [Accessed: February 22, 2019]

[32] Yu W, Xie H, Wang X, Wang X. Significant thermal conductivity enhancement for nanofluids containing graphene nanosheets. Physics Letters A. 2011;**375**(10):1323-1328

[33] Vinoth R, Ganesh Babu S, Bharti V, Gupta V, Navaneethan M, Venkataprasad Bhat S, et al. Ruthenium based metallopolymer grafted reduced graphene oxide as a new hybrid solar light harvester in polymer solar cells. Scientific Reports. 2017;**7**:43133

[34] Priyadarsini S, Mohanty S, Mukherjee S, Basu S, Mishra M. Graphene and graphene oxide as nanomaterials for medicine and biology application. Journal of Nanostructure in Chemistry. 2018;**8**:123-137

Chapter 2

Water Remediation by G-/GO-Based Photocatalysts

Humaira Seema

Abstract

Graphene, a two-dimensional sheet of sp^2 hybridized carbon atoms, has shown to be the most fascinating and promising option among nanomaterials for a variety of applications, because of its unique structure and tunable physiochemical properties. It can be either in the pure form or in its modified derivatives that include graphene oxide (GO), reduced graphene oxide (rGO), graphene-metal nanoparticle composites, graphene-polymer hybrids, and graphene/organic structures that showed improved results while maintaining inherent properties of the material. These modified nanostructures have a variety of applications as catalysts, energy storage/conversion, antimicrobial, and water decontaminant. In the field of environmental science, graphene has been widely used for molecular sieving involving gas phase separation and organic waste removal from water, due to its biocompatibility, various functional groups, and accessible surface area. Modified graphene can also serve as a semiconductor that can increase the efficiency of the photocatalytic ecosystems that results in the inactivation of the microorganisms causing the organic chemicals to degrade.

Keywords: graphene, environmental, remediation, photocatalyst, water

1. Introduction

Recently photocatalysis by using semiconductors has fascinated universal consideration for its energy-related and environmental applications. Nevertheless, the decrease in the efficiency of the photocatalysis restricted its practical applications because of the prompt reunion of photogenerated electrons and holes. Thus, to decrease the reunion of charge carriers is significant for improvement of semiconductor photocatalysis. Among numerous approaches, water remediation has been done by rGO-/GO-based materials which are the most favorable candidates due to their high capacity of dye adsorption, prolonged light absorption range, improved separation of charge carriers, and transportation properties leading to improved photoconversion efficiency of the photocatalytic materials [1–74].

1.1 Graphene (rGO)-based photocatalysts

Various numbers of graphene-based photocatalysts have been prepared with its derivatives which mainly comprise metal oxides (e.g., P25 [1, 8], TiO_2 [9–34, 36, 37], ZnO [17, 39–43], CuO [44], SnO_2 [13, 45], WO_3 [46]), metals (e.g., Cu [51], Au [52]), metal-metal oxides (e.g., Ag-TiO_2 [35]), upconversion material—P25 (e.g., YF_3:Yb^{3+},Tm^{3+}—TiO_2 [38]), salts (e.g., CdS [47–49], ZnS [50], $ZnFe_2O_4$ [53],

$MnFe_2O_4$ [54], $NiFe_2O_4$ [55], $CoFe_2O_4$ [56], Bi_2WO_6 [57–59], Bi_2MoO_6 [60], $InNbO_4$ [61], ZnSe [63]), Ag/AgCl [62]), and other carbon material (e.g., CNT [64]).

1.2 Graphene oxide (GO)-based photocatalysts

Graphene oxide (GO) has recently received considerable attention due to oxygen-containing functional groups which increase its solubility in solvents for the preparation of GO-based nanocomposites required for photodegradation of pollutants [65–74]. GO-based nanocomposites mainly include metal oxides (TiO_2) [66–72], metal-free polymers [73], and silver/silver halides [74].

2. Preparation of rGO-/GO-based composite photocatalysts

Some of the commonly used synthesis techniques include in situ growth strategy, solution mixing, hydrothermal/ solvothermal, and microwave-assisted process.

2.1 In situ growth strategy

This method is usually used to prepare reduced graphene oxide-/graphene oxide-based metal composites. Zhang et al. reported that TiO_2/graphene composite photocatalyst [14] is synthesized by a simple liquid-phase deposition technique. Moreover, adopting a similar approach, Wang et al. prepared nanocarbon/TiO_2 nanocomposites where titania nanoparticles were decorated by thermal reaction on the surfaces of three different dimensional nanocarbons [9]. While in thermal reduction method, TiO_2/graphene composite [12] with a remarkable visible light photocatalytic activity was prepared by Zhang et al. using a heat treatment method of GO, where GO changed to reduced graphene oxide. Uniform ZnO nanoparticles were found on functionalized graphene sheets evenly via thermal decay of mixture of zinc salt, graphene oxide, and poly(vinyl pyrrolidone) [39].

Furthermore, Sn^{2+} or Ti^{3+} ions were converted to oxides at low temperatures, while GO was reduced to reduced graphene oxide by tin or titanium salts in redox method [13–45]. In our recent work, we prepared SnO_2-G nanocomposite which displayed higher photocatalytic activity in sunlight as compared to bare metal oxide nanoparticles as shown in **Figure 1** [45]. Similarly reduced graphene oxide-zinc oxide composite was prepared where zinc ions were decorated on GO sheets and transformed to metal oxide nanoparticles by using chemical reagents at 150°C. Reduced graphene oxide-ZnO photocatalyst is formed by reducing the graphene oxide [43].

Li et al. prepared uniform mesoporous titania nanospheres on reduced graphene oxide layers via a process of a template-free self-assembly [20]. Du et al. [21] also developed the macro-mesoporous titania-reduced graphene oxide composite film by a confinement of a self-assembly process as shown in **Figure 2**.

Moreover Kim et al. synthesized strongly coupled nanocomposites of layered titanate and graphene by electrostatically derived self-assembly between negatively charged G nanosheets and positively charged TiO_2 nanosols, followed by a phase transition of the anatase TiO_2 component into layered titanate [37]. Chen et al. prepared graphene oxide/titania composites by using the self-assembly technique [72].

While Cu ion-modified reduced graphene oxide [51] prepared by an immersion technique displayed a high photocatalytic activity, gold nanoparticles were decorated on the surface of the reduced graphene oxide through spontaneous chemical reduction of $HAuCl_4$ by GOR [52] as shown in **Figure 3**.

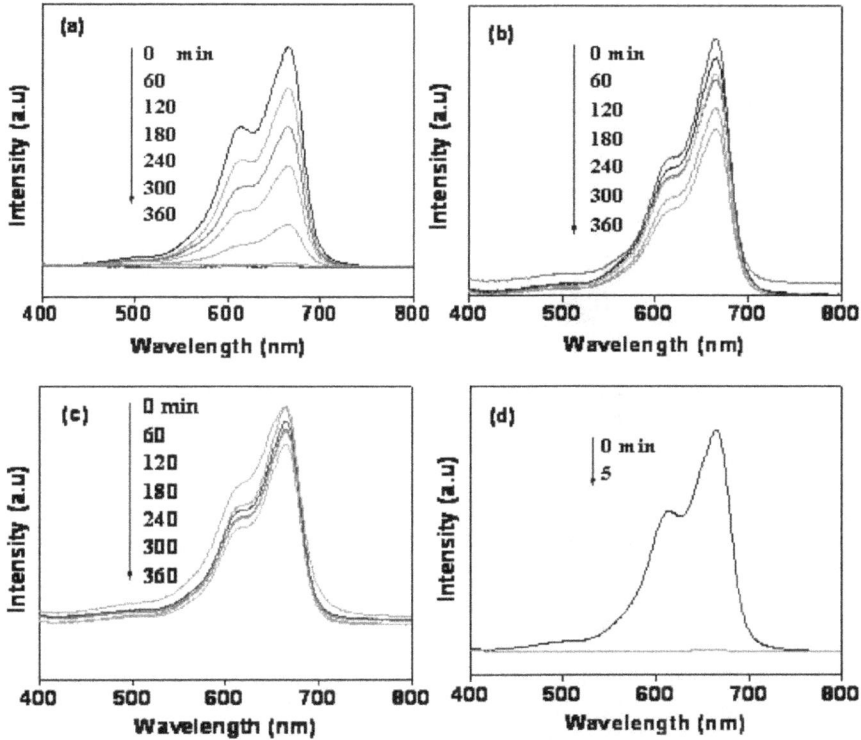

Figure 1.
Time-dependent absorption spectra of MB solution during UV light irradiation in the presence of (a) SnO$_2$ and (b) reduced graphene oxide-SnO$_2$ and during sunlight irradiation in the presence of (c) SnO$_2$ and (d) reduced graphene oxide-SnO$_2$. Reprinted with permission of the publisher [45].

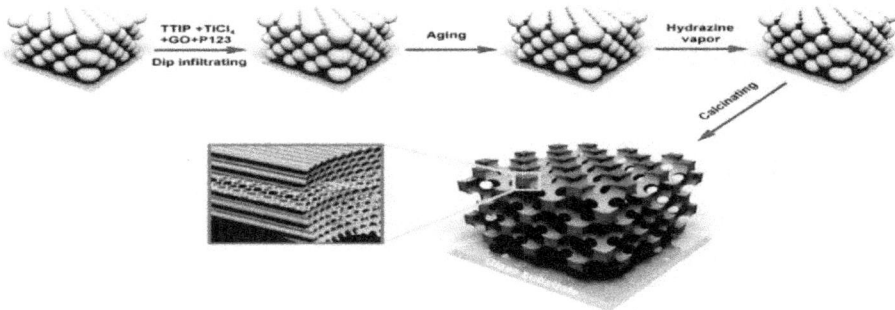

Figure 2.
Schematic view for the preparation of a macro-mesoporous TiO$_2$-reduced graphene oxide composite film. Reprinted with permission of the publisher [21].

Bi$_2$WO$_6$/reduced graphene oxide photocatalysts were successfully prepared via in situ refluxing method in the presence of GO [57]. Zhang et al. presented reduced graphene oxide sheet grafted Ag@AgCl plasmonic photocatalyst with high activity via a precipitation reaction followed by reduction [62]. TiO$_2$-GO was well prepared at 80°C by using GO and titanium sulfate as precursors [66].

Liu et al. have established a process of water/toluene two-phase for self-assembling TiO$_2$ nanorods on graphene oxide [69, 70]. Jiang et al. prepared GO/titania

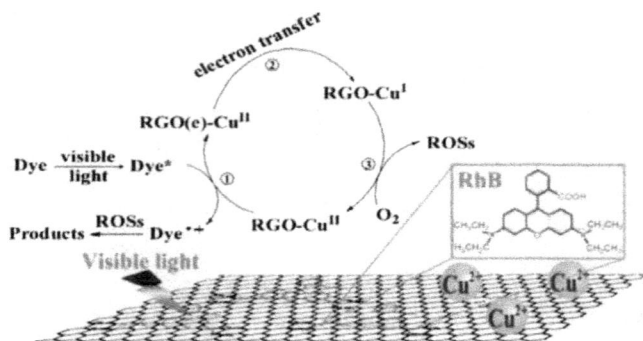

Figure 3.
Possible mechanism of photosensitized degradation of dyes over a rGO Cu composite under visible light irradiation. Reprinted with permission of the publisher [52].

composite by in situ depositing titania on GO through liquid-phase deposition, followed by a calcination at 200°C [71].

GO nanostructures are prepared by modified Hummer's method, which has promising applications in photocatalysis [65].

2.2 Solution mixing method

It has been widely used to prepare graphene-based photocatalysts. Previously, titania nanoparticles and GO colloids have been mixed by ultrasonication followed by ultraviolet (UV)-assisted photocatalytic reduction of GO to yield graphene-titania nanocomposites [18, 23, 31].

Akhavan and Ghaderi used a similar strategy to prepare the titania/reduced graphene oxide composite thin film [25].

Guo et al. [28] prepared TiO_2/graphene composite via sonochemical method. GO/g-C_3N_4 with efficient photocatalytic capability was also fabricated by the same sonochemical approach [73].

ZnO and GO mixture was dispersed by ultrasonication followed by chemical reduction of GO to graphene ultimately leading to synthesize ZnO/graphene composite [40]. The G-hierarchical ZnO hollow sphere composites are synthesized by Luo et al. by using a simple ultrasonic treatment of the solution [43].

Cheng et al. [40] presented a new facile ultrasonic approach to prepare graphene quantum dots (GQDs), which exhibited photoluminescent in a water solution. The water/oil system is used by Zhu et al. [74] to produce graphene oxide enwrapped Ag/AgX (X = Br, Cl) composites. Graphene oxide and silver nitrate solution were added to chloroform solution of surfactants stirring condition at room temperature to produce hybrid composites which displayed high photocatalytic activity under visible light irradiation as shown in **Figure 4**. Titania/graphene oxide composites were synthesized using one-step colloidal blending method [68].

2.3 Hydrothermal/solvothermal method

This one-pot process can lead to highly crystalline nanostructures, which operates at elevated temperatures in an autoclave to generate high pressure, without calcination, and at the same time GO reduced to rGO. Typically, graphene-based composites, e.g., P25 [1, 8], TiO_2 [15, 16, 24, 29, 30, 32–34], Ag-TiO_2 [35], UC-P25 [38], WO_3 [46], CdS [49], $ZnFe_2O_4$ [53], $MnFe_2O_4$ [54], $NiFe_2O_4$ [55], Bi_2WO_6 [58, 59], Bi_2MoO_6 [60], $InNbO_4$ [61], and ZnSe [63], have been prepared by the

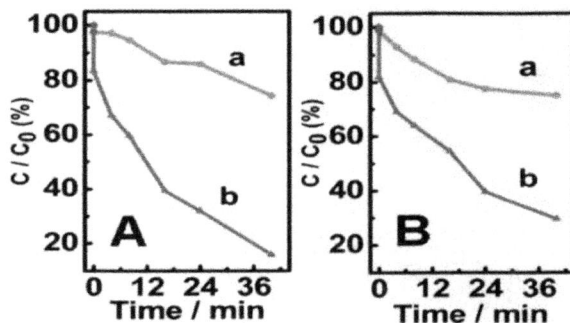

Figure 4.
(A) Photocatalytic activities of silver/silver bromide (a) and silver/silver bromide/GO (b) nanospecies for photodegradation of MO molecules under visible light irradiation and (B) those of the Ag/AgCl (a) and Ag/AgCl/GO (b) nanospecies. Reprinted with permission of the publisher [74].

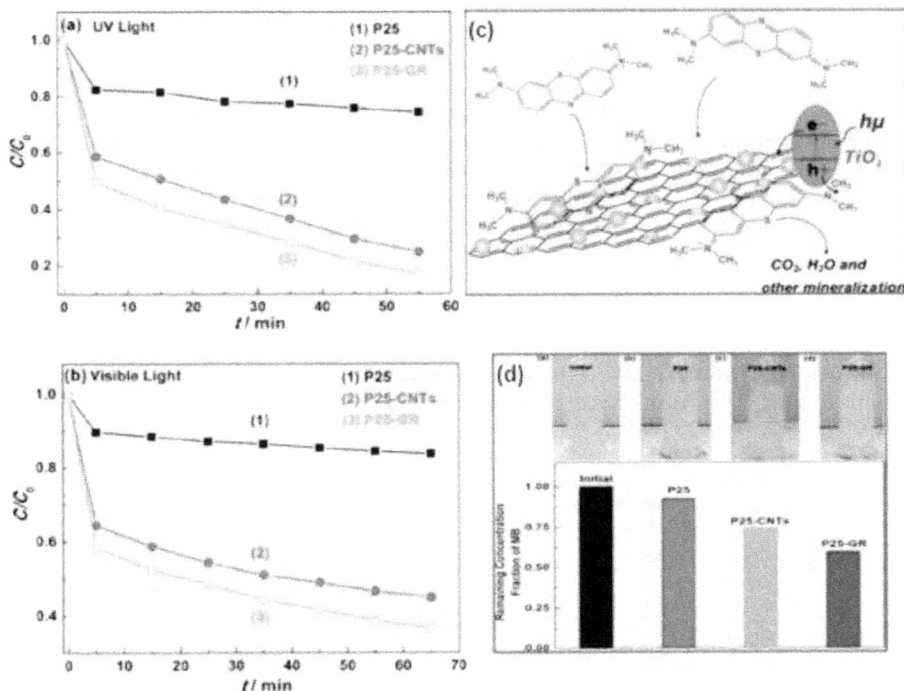

Figure 5.
Photodegradation of MB under (a) UV light (λ = 365 nm) and (b) visible light (λ > 400 nm) over (1) P25, (2) P25-CNTs, and (3) P25-GR photocatalysts, respectively. (c) Schematic structure of P25-GR and process of the photodegradation of MB over P25-GR. (d) Bar plot showing the remaining MB in solution: (1) initial and equilibrated with (2) P25, (3) P25-CNTs, and (4) P25-GR in the dark after 10-min stirring. Pictures of the corresponding dye solutions are on the top for each sample. Reprinted with permission of the publisher [8].

hydrothermal process, while others such as TiO_2 [11, 22, 26, 27], CuO [44], CdS [48], and $CoFe_2O_4$ [56] are prepared by the solvothermal process.

Li et al. have prepared P25-G nanocomposite using GO and P25 as raw materials via hydrothermal technique [8]. As illustrated in **Figure 5**, the photocatalysis determines that composite showed improved activity toward the photodegradation of methylene blue (MB).

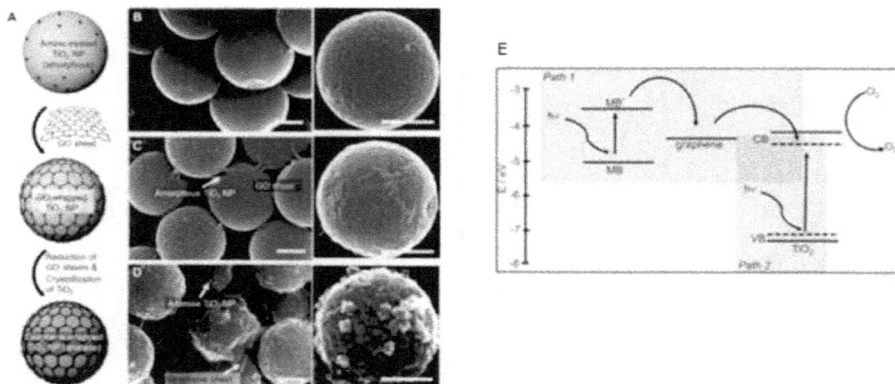

Figure 6.
(A) Schematic illustration of synthesis steps for graphene-wrapped anatase TiO$_2$ nanoparticles (NPs) and corresponding SEM images of (B) bare amorphous TiO$_2$ NPs, (C) GO-wrapped amorphous TiO$_2$ NPs, and (D) graphene-wrapped anatase TiO$_2$ NPs (scale bar: 200 nm); (E) the suggested mechanism for the photocatalytic degradation of MB by graphene-wrapped anatase TiO$_2$ NPs under visible light irradiation. Reprinted with permission of the publisher [8].

Lee et al. synthesized graphene oxide (GO)-wrapped TiO$_2$ nanoparticles by combining positively charged TiO$_2$ nanoparticles with negatively charged GO nanosheets, as shown in SEM images in **Figure 6**. Furthermore, it demonstrates the reduction of graphene oxide to reduced graphene oxide and the crystallization of amorphous titania nanoparticles which occurred after a hydrothermal treatment.

2.4 Microwave-assisted method

In situ microwave irradiation is a facile method which has been used for the simultaneous formation of metal oxide (e.g., TiO$_2$ [17], ZnO [17, 41], CdS [47],

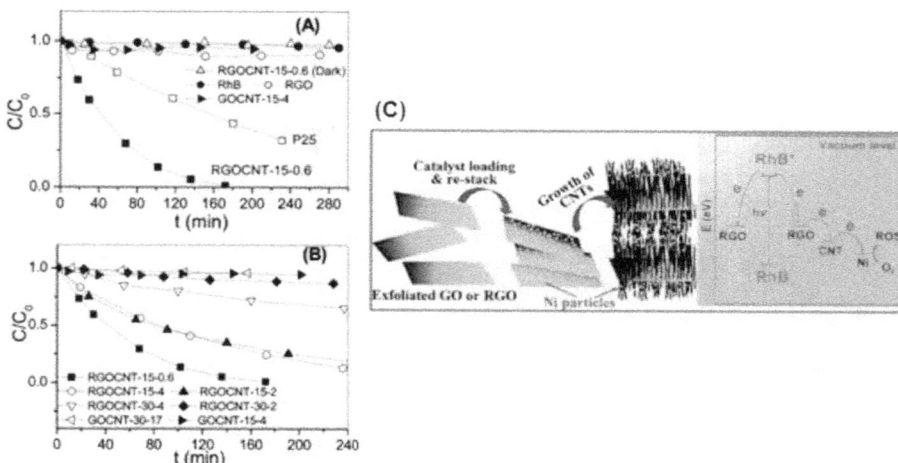

Figure 7.
(A) Photocatalytic degradation for RhB under different experimental conditions with catalysts GOCNT-15-4 and P25. (B) Photocatalytic properties of different samples in degrading RhB. (C) Experimental steps of pillaring GO and RGO platelets with CNTs while energy diagram showing the proposed mechanism of photosensitized degradation of RhB under visible light irradiation. Reprinted with permission of the publisher [64].

ZnS [50]) and reduction of GO. The drawback of this process is that it did not show its fine control over the uniform size and surface distribution of nanoparticles on G surfaces.

2.5 Other methods

In addition to the abovementioned examples, graphene-based photocatalysts are synthesized by developing new synthetic strategies, e.g., electrospinning [10] and chemical vapor deposition (CVD) [64].

Zhao et al. pillared reduced graphene oxide platelets with carbon nanotubes using the CVD method with acetonitrile as the carbon source and nickel nanoparticles as the catalysts as shown in **Figure 7**.

Photocatalytic TiO_2 films were prepared by Yoo et al. using RF magnetron sputtering and GO solutions with different concentrations of GO in ethanol which were coated on TiO_2 films [67]. Graphene film was formed on the surface of TiO_2 nanotube arrays through in situ electrochemical reduction of GO dispersion by cyclic voltammetry [19].

3. Photocatalysis

Due to widespread environmental applications, photocatalysis has fascinated an increasing consideration. The graphene-/graphene oxide-based photocatalyst revealed a significant improvement of photocatalytic degradation of methylene blue (MB) [1, 8, 11, 12, 15, 18, 21, 22, 26, 28, 30–33, 35, 36, 40, 41, 43, 45, 48, 50, 52–56, 60, 61, 67–69], rhodamine B (RhB) [13, 20, 24, 27, 32, 42, 44, 51, 52, 56–59, 62, 64, 73], methyl orange (MO) [9, 10, 14, 37, 38, 49, 63, 66, 71, 72, 74], anthracene-9-carboxylic acid (9-AnCOOH) [19], phenol [22, 54], 2,4-dichlorophenoxyacetic acid (2,4-D) [23], 2,4-dichlorophenol [61, 73], malachite green (MG) [29], 2-propanol [34], rhodamine 6G (Rh 6G) [39], rhodamine B 6G (RhB 6G) [46], orange ll [52], 2,4-dichlorophenol (2,4-DCP) [61], acid orange 7(AO 7) [64], and resazurin (RZ) [65], as well as photocatalytic reduction of Cr(VI) [17, 47, 71], along with photocatalytic antibacterial activity for killing *E. coli* bacteria [25] by UV [1, 8–10, 13, 14, 16–24, 26, 28–34, 37, 39–45, 50, 54, 65–71], as well as visible irradiation [1, 8–13, 15, 16, 20, 22, 25, 27, 30–33, 35–38, 42, 45–49, 51–64, 67, 68, 72–74], in water which are briefly summarized in **Table 1**.

Photocatalysts	Mass fraction	Preparation strategy	Photocatalytic experiments	Performances as compared to reference photocatalyst	Type of irradiation	References
(1) rGO-based						
P25-rGO	0.2% G	Hydrothermal method	Photodegradation of MB	1.17 times higher than P25; DP of 60%	UV	1
	5%			1.50 times		
	30%			0.97 times		
	0.2%			1.42 times higher than P25; DP of 28%	Visible	
	5%			2.32 times		
	30%			0.75 times		
P25-rGO	1.0% G	Hydrothermal method	Photodegradation of MB	3.40 or 1.21 times higher than P25 or P25-CNTs; DP of 25% or 70%, respectively (2% = 90 min)	UV	2

Photocatalysts	Mass fraction	Preparation strategy	Photocatalytic experiments	Performances as compared to reference photocatalyst	Type of irradiation	References
				4.33 or 1.18 times higher than P25 or P25-CNTs; DP of 15% or 55%, respectively (2% = 90 min)	Visible	
TiO$_2$-rGO	10 mg G	In situ growth strategy (thermal treatment)	Degradation of MO	2.05 times higher than P25; DP of 40%	UV	3
				5.46 times higher than P25; 15%	Visible	
TiO$_2$-rGO	0.75% G	Electrospin-ning method	Degradation of MO	1.51 times higher than TiO$_2$; DP of 54%	UV	4
				2.04 times higher than TiO$_2$; DP of approx. 22%	Visible	
TiO$_2$-rGO	10 mg G	Solvothermal method	Photodegradation of MB	2.32 or 1.50 times higher than pure TiO$_2$ or P25; DP of 25% or 39%, respectively	Visible	5
	30 mg			3.0 or 1.92 times		
	50 mg			2.88 or 1.84 times		
TiO$_2$-rGO	10 mg G	In situ growth strategy (thermal reduction method)	Photodegradation of MB	7.0 times higher than pure P25; DP of 10%	Visible	6
TiO$_2$-rGO	No data	In situ growth strategy (redox method)	Photodegradation of RhB	1.16 times higher than P25 reaction rate constant = 0.0049 min^{-1}	Visible	7
				0.53 times higher than P25 reaction rate constant = 0.043 min^{-1}	UV	
SnO$_2$-rGO				2.24 times higher than P25 reaction rate constant = 0.0049 min^{-1}	Visible	
				0.62 times higher than P25 reaction rate constant = 0.043 min^{-1}	UV	
TiO$_2$-rGO	20 mg G	In situ growth strategy (simple liquid-phase deposition method)	Photodegradation of MO	1.89 times higher than P25 and graphene; DP of 45%	UV	8
TiO$_2$-rGO	No data	Hydrothermal method	Photodegradation of MB	13.04 or 10.62 times higher than P25 or anatase TiO$_2$; reaction rate constant = 0.0026 min^{-1} or 0.0032 min^{-1}, respectively	Visible	9
TiO$_2$-rGO	20:1	Hydrothermal method	Photodegradation of RhB	1.63 times higher than P25; DP of 52%	UV	10

Photocatalysts	Mass fraction	Preparation strategy	Photocatalytic experiments	Performances as compared to reference photocatalyst	Type of irradiation	References
				3.33 times higher than P25; DP of 15%	Visible	
TiO_2-rGO	0.8% G	Microwave-assisted method	Photocatalytic reduction of Cr(VI)	1.09 or 1.30 times higher than pure TiO_2 or commercial P25 = removal rate of 83% or 70%, respectively	UV	11
rGO-w-TiO_2	1:10	Solution mixing method	Photodegradation of MB	1.25 times higher than P25; DP of 80%	UV	12
TiO_2-rGO film	No data	Cyclic voltammetric reduction method	Photodegradation of anthracene-9-carboxylic acid (9-AnCOOH)	2.13 times higher than bare TiO_2 nanotubes; DP of 46%	UV	13
TiO_2-rGO	6.5% G	In situ growth strategy (self-assembly synthesis)	Photodegradation of RhB	3.92 times higher than TiO_2; DP of 25%	UV-vis	14
TiO_2-rGO	0.6% G	In situ growth strategy (self-assembly method)	Photodegradation of MB	1.57 times higher than TiO_2; reaction rate constant = 0.045 min^{-1}	UV	15
TiO_2-rGO	No data	Solvothermal method	Photodegradation of phenol	1.68 times higher than P25; DP of 48%	UV	16
				3.10 times higher than P25; DP of 20%	Visible	
			Photodegradation of MB	3.5 times higher than P25; DP of 20%	Visible	
TiO_2-rGO film	No data	Solution mixing method	Photodegradation of 2,4-dichlor-ophen-oxyacetic acid (2,4-D)	4.0 times higher than TiO_2 film; reaction rate constant = 0.002 min^{-1}	UV	17
TiO_2-rGO	10% GO	Hydrothermal method	Photodegradation of RhB	4.0 or 2.94 times higher than pure TiO_2 or P25; reaction rate constant = 0.05 or 0.068 min^{-1}	UV	18
TiO_2-rGO	No data	Solution mixing method	Photocatalytic antibacterial activity for killing *E. coli* bacteria	7.55 times higher than TiO_2; reaction rate constant = 0.0086 min^{-1}	Visible	19
TiO_2-rGO	0.3 mg GO	Solvothermal method	Photodegradation of MB	2.08 times higher than P25; DP of 40.8%	UV	20
TiO_2-rGO	No data	Solvothermal method	Photodegradation of RhB	2.79 times higher than P25; reaction rate constant = 0.0162 min^{-1}	Visible	21
TiO_2-rGO	75% G	Sonochemical method	Photodegradation of MB	2.57 times higher than P25; reaction rate constant = 0.0054 min^{-1}	UV	22
TiO_2-rGO	10% G	Hydrothermal method	Photodegradation of Malachite green	3.09 times higher than TiO_2 nanotubes; reaction rate constant = 0.0218 min^{-1}	UV	23

Photocatalysts	Mass fraction	Preparation strategy	Photocatalytic experiments	Performances as compared to reference photocatalyst	Type of irradiation	References
TiO$_2$-rGO	No data	Hydrothermal method	Photodegradation of MB	1.46 times higher than P25; DP of 65%	UV	24
				2.41 times higher than P25; DP of 29%	Visible	
rGO @TiO$_2$	1:3	Solution mixing method	Photodegradation of MB	4.0 or 1.73 times higher than P25 or physical mixture of G-P25 (1:3); DP of 13% or 30%, respectively	Visible	25
			Photodegradation of MB	2.93–2.20 times higher than P25 or physical mixture of G-P25 (1:3); DP of 30–40%	UV	
TiO$_2$-B-doped rGO	2 mg G	Hydrothermal method	Photodegradation of MB	4.30 times higher than TiO$_2$; reaction rate constant = 0.010 min^{-1}	UV-vis	26
			Photodegradation of RhB	1.6 times higher than TiO$_2$; reaction rate constant = 0.005 min^{-1}		
TiO$_2$-N-doped rGO			Photodegradation of MB	2.4 times higher than TiO$_2$; reaction rate constant = 0.010 min^{-1}		
			Photodegradation of RhB	3.2 times higher than TiO$_2$; reaction rate constant = 0.005 min^{-1}		
TiO$_2$-rGO-TiO$_2$	0.01 g G	Hydrothermal method	Photodegradation of MB	4 times higher than TiO$_2$	UV-vis	27
TiO$_2$-rGO/ MCM-41	0.05% G	Hydrothermal method and Thermal method	Photodegradation of 2-propanol	1.4 times higher than TiO$_2$/MCM-41; conversion rate of 26%	UV	28
	0.15%			1.7 times		
	0.4%			1.27 times		
	0.6%			0.96 times		
Ag-TiO$_2$-rGO	No data	Hydrothermal and solution mixing method	Photodegradation of MB	Enhancement	Visible	29
RutileTiO$_2$-GQD/anatase TiO$_2$-GQD	0.05 g G	Solution mixing method	Degradation of MB	Enhancement for rutile TiO$_2$/GQD than anatase TiO$_2$/GQD	Visible	30
Layered titanate rGO	No data	In situ growth strategy (self-assembly method)	Photodegradation of MO	Enhancement as compared to bulk-layered titanates or nanocrystalline-layered titanate	UV-vis	31
UC-P25-rGO UC = YF$_3$:Yb^{3+},Tm^{3+}	4 mg GO	Hydrothermal method	Photodegradation of MO	2.88 or times higher than P25 or P25-G or UC-P25; DP of 27% or 53% or 46%, respectively	Visible	32
ZnO-rGO	0.6% G	Microwave-assisted method	Photocatalytic reduction of Cr(VI)	1.12 or 0.92 times higher than pure ZnO or P25; removal rate of 58 or 70%, respectively	UV	33

Photocatalysts	Mass fraction	Preparation strategy	Photocatalytic experiments	Performances as compared to reference photocatalyst	Type of irradiation	References
	0.8% G			1.46 or 1.21 times		
	1.0% G			1.68 or 1.40 times		
ZnO-FGS	0.1 g GO	In situ growth strategy (thermal method)	Photodegradation of Rh 6G	Enhancement	UV	34
ZnO-rGO	0.1% G	Solution mixing method (sonochemical)	Photodegradation of MB	2.13 times higher than ZnO; reaction rate constant = 0.022 min^{-1}	UV	35
	0.5%			2.54 times		
	1.0%			3.13 times		
	2.0%			4.45 times		
	3.0%			4.13 times		
	5.0%			3.27 times		
ZnO-rGO	1.1% G	Microwave-assisted method	Photodegradation of MB	1.29 times higher than ZnO; DP of 68%	UV	36
ZnO@ rGO		In situ growth strategy (chemical deposition method)	Photodegradation of RhB	1.05 times higher than ZnO; DP of 95%	UV	37
				1.02 times higher than ZnO; DP of 98%	Visible	
ZnO-rGO	3.56% G	Solution mixing method (ultrasonic method)	Photodegradation of MB	2.25 times higher than ZnO; DP of 40%	UV	38
CuO-rGO	No data	Solvothermal method	Photodegradation of RhB in the presence of H$_2$O$_2$	2.50 times higher than ZnO; DP of 40%	UV	39
SnO$_2$-rGO	5% G	In situ growth strategy (redox method)	Photodegradation of MB	0.40 or times higher than SnO$_2$; DP of 100%	UV	40
				24.86 times higher than SnO$_2$; DP of 4%	Visible	
WO$_3$-rGO	3.5% G	Hydrothermal method	Photodegradation of RhB 6G	2.2 or 53 times higher than WO$_3$ nanorods or WO$_3$ particles; reaction rate constant = 0.00167or 0.000069 min^{-1}, respectively	Visible	41
CdS-rGO	1.5% G	Microwave-assisted method	Photocatalytic reduction of Cr(VI)	1.16 times higher than CdS = removal rate of 79%	Visible	42
CdS-rGO	5% G	Solvothermal method	Photodegradation of MB	2.5 times higher than CdS; DP of 37.6%	Visible	43
CdS-rGO	0.01:1	Hydrothermal method	Photodegradation of MO	7.86 times higher than CdS; reaction rate constant = 0.0075 min^{-1}	Visible	44

Photocatalysts	Mass fraction	Preparation strategy	Photocatalytic experiments	Performances as compared to reference photocatalyst	Type of irradiation	References
ZnS-rGO	No data	Microwave-assisted method	Photodegradation of MB	4 times higher than P25; DP of 25%	UV	45
Cu-rGO	No data	In situ growth strategy (immersion method)	Photodegradation of RhB	2.94 or 30.61 times higher than P25 or graphene; reaction rate constant = 0.0051 min^{-1} or 0.00049 min^{-1}, respectively	Visible	46
Au-rGO	No data	In situ growth strategy (chemical reduction)	Photodegradation of RhB	1.77 times higher than P25; reaction rate constant = 0.0049 min^{-1}	Visible	47
			Photodegradation of MB	8.36 times		
			Photodegradation of orange II	0.19 times		
$ZnFe_2O_4$-rGO	20% G	Hydrothermal method	Photodegradation of MB in the presence of H_2O_2	4.50 times higher than $ZnFe_2O_4$ (DP of 22% = 90 min)	Visible	48
$MnFe_2O_4$-rGO	30% G	Hydrothermal method	Photodegradation of MB	9.62 times higher than $MnFe_2O_4$; DP of 10%	Visible	49
			Photodegradation of MB	1.33 times higher than $MnFe_2O_4$; DP of 75%	UV	
			Photodegradation of phenol	1.13 times higher than $MnFe_2O_4$; DP of 75%	UV	
$NiFe_2O_4$-rGO	25% G	Hydrothermal method	Photodegradation of MB	Enhancement as compared to $NiFe_2O_4$; reaction rate constant almost zero (no photocatalytic activity)	Visible	50
$CoFe_2O_4$-rGO	No data	Solvothermal method	Photodegradation of RhB and MB	Enhancement	Visible	51
Bi_2WO_6-rGO	1% G	In situ growth strategy (refluxing method)	Photodegradation of RhB	1.30 times higher than Bi_2WO_6; DP of 50%	Visible	52
	2.5%			1.40 times		
	5%			1.80 times		
	10%			1.10 times		
	15%			0.80 times		
Bi_2WO_6-rGO	1% G	Hydrothermal method	Photodegradation of RhB	Enhancement as compared to Bi_2WO_6	Visible	53
Bi_2WO_6-rGO	No data	Hydrothermal method	Photodegradation of RhB	2.04 times higher than Bi_2WO_6; DP of 44% in 4 min	Visible	54
Bi_2MoO_6-rGO	0.5% G	Hydrothermal method	Photodegradation of MB	2.45 times higher than pure Bi_2MoO_6; reaction rate constant 0.0037 min^{-1}	Visible	55
	1%			3.67 times		

Photocatalysts	Mass fraction	Preparation strategy	Photocatalytic experiments	Performances as compared to reference photocatalyst	Type of irradiation	References
InNbO$_4$-rGO	No data	Hydrothermal method	Photodegradation of MB	1.87 times higher than InNbO$_4$; reaction rate constant = 0.0185 min^{-1}	Visible	56
			Photodegra-dation of 2,4–dichloro-phenol	2.10 times higher than InNbO$_4$ reaction rate constant = 0.0256 min^{-1}		
Ag@AgCl-rGO	0.22% G	Solution mixing method	Photodegradation of RhB	3.88 times higher than Ag@AgCl reaction rate constant = 0.060 min^{-1}	Visible	57
	0.44%			4.55 times		
	1.56%			5.1 times		
ZnSe-N-doped rGO	18 mg G	Hydrothermal method	Photodegradation of MO	Enhancement as compared to ZnSe; (no photocatalytic activity)	Visible	58
CNT-rGO	No data	Chemical vapor deposition (CVD) method	Photodegradation of RhB	4.28 times higher than P25; reaction rate constant = 0.0049 min^{-1}	Visible	59
(2) GO-based						
GO	1 mg GO	Solution mixing method (modified Hummers' method)	Photocatalytic reduction of resazurin (RZ)	No data	UV	60
TiO$_2$-GO	No data	In situ growth strategy	Photodegradation of MO	2.27 times higher than pure P25; DP of 38.4%	UV	61
TiO$_2$-GO	0.03 mg GO	RF magnetron sputtering followed by coating	Photodegradation of MB	2.5 times higher than TiO$_2$; DP of 20%	UV	62
				1.75 times	Visible	
TiO$_2$-GO	1.2% GO	Solution mixing method (simple colloidal blending method)	Photodegradation of MB	4.51 times higher than P25 reaction rate constant = 0.0084 min^{-1}	UV	63
	4.3%			4.98 times		
	8.2%			8.59 times		
	1.2%			1.36 times higher than P25 reaction rate constant = 0.0033 min^{-1}	Visible	
	4.3%			3.03 times		
	8.2%			7.15 times		
TiO$_2$-GO	50 mg GO	In situ growth strategy	Photodegradation of MB	1.41 times higher than P25; DP of 70%	UV	64
TiO$_2$-GO	500 mg GO	In situ growth strategy (two phase assembling method)	Photodegradation of acid orange 7 (AO 7)	11.59 times higher than P25 reaction rate constant = 0.0182 min^{-1}	UV	65

Photocatalysts	Mass fraction	Preparation strategy	Photocatalytic experiments	Performances as compared to reference photocatalyst	Type of irradiation	References
TiO$_2$-GO	No data	In situ growth strategy (thermal treatment method)	Photodegradation of MO	7.44 times higher than P25; reaction rate constant = 0.0426 min^{-1}	UV	66
			Photocatalytic reduction of Cr(VI)	5.44 times higher than P25; conversion rate = 0.0127 min^{-1}		
TiO$_2$-GO	0.13% C element	In situ growth strategy (self-assembly method)	Photodegradation of MO	1.18 times higher than pure P25; DP of 22%	Visible	67
	0.14%			1.59 times		
	0.25%			1.0 times		
	0.51%			0.82 times		
g-C$_3$N$_4$-GO	1 g GO	Solution mixing method (sonochemical method)	Photodegradation of RhB and 2,4-dichloro-phenol	1.90 times higher than g-C$_3$N$_4$; DP of 49.5%	Visible	68
Ag/AgCl/GO	No data	Solution mixing method (surfactant-assisted assembly protocol via an oil/water microemulsion)	Photodegradation of MO	2.84 times higher than Ag/AgCl; DP of 25%	Visible	69
Ag/AgBr/GO			Photodegradation of MO	3.40 times higher than Ag/AgBr; DP of 25%	Visible	

Table 1.
Photocatalytic degradation of pollutants.

Author details

Humaira Seema
Institute of Chemical Sciences, University of Peshawar, Pakistan

*Address all correspondence to: hawkkhan2@gmail.com

IntechOpen

References

[1] Zhang Y, Tang ZR, Fu X, Xu YJ. TiO$_2$-graphene nanocomposites for gas-phase photocatalytic degradation of volatile aromatic pollutant: Is TiO$_2$-graphene truly different from other TiO$_2$-carbon composite materials? ACS Nano. 2010;**4**:7303-7314

[2] Huang X, Qi X, Boey F, Zhang H. Graphene-based composites. Chemical Society Reviews. 2012;**41**:666-686

[3] Han L, Wang P, Dong S. Progress in graphene-based photoactive nanocomposites as a promising class of photocatalyst. Nanoscale. 2012;**4**:5814-5825

[4] Xiang Q, Yu J, Jaroniec M. Graphene-based semiconductor photocatalysts. Chemical Society Reviews. 2012;**41**:782-796

[5] Zhang N, Zhang Y, Xu YJ. Recent progress on graphene-based photocatalysts: Current status and future perspectives. Nanoscale. 2012;**4**:5792-5813

[6] Le NH, Seema H, Kemp KC, Ahmed N, Tiwari JN, Park S, et al. Solution-processable conductive micro-hydrogels of nanoparticle/graphene platelets produced by reversible self-assembly and aqueous exfoliation. Journal of Materials Chemistry A. 2013;**1**:12900-12908

[7] Kemp KC, Seema H, Saleh M, Le NH, Mahesh K, Chandra V, et al. Environmental applications using graphene composites: Water remediation and gas adsorption. Nanoscale. 2013;**5**:3149-3171

[8] Zhang H, Lv X, Li Y, Wang Y, Li J. P25-graphene composite as a high performance photocatalyst. ACS Nano. 2009;**4**:380-386

[9] Wang F, Zhang K. Physicochemical and photocatalytic activities of self-assembling TiO$_2$ nanoparticles on nanocarbons surface. Current Applied Physics. 2012;**12**:346-352

[10] Zhu P, Nair AS, Shengjie P, Shengyuan Y, Ramakrishna S. Facile fabrication of TiO$_2$-graphene composite with enhanced photovoltaic and photocatalytic properties by electrospinning. ACS Applied Materials & Interfaces. 2012;**4**:581-585

[11] Zhou K, Zhu Y, Yang X, Jiang X, Li C. Preparation of graphene-TiO$_2$ composites with enhanced photocatalytic activity. New Journal of Chemistry. 2011;**35**:353-359

[12] Zhang Y, Pan C. TiO$_2$/graphene composite from thermal reaction of graphene oxide and its photocatalytic activity in visible light. Journal of Materials Science. 2011;**46**:2622-2626

[13] Zhang J, Xiong Z, Zhao XS. Graphene-metal-oxide composites for the degradation of dyes under visible light irradiation. Journal of Materials Chemistry. 2011;**21**:3634-3640

[14] Zhang H, Xu P, Du G, Chen Z, Oh K, Pan D, et al. A facile one-step synthesis of TiO$_2$/graphene composites for photodegradation of methyl orange. Nano Research. 2011;**4**:274-283

[15] Lee JS, You KH, Park CB. Highly photoactive, low bandgap TiO$_2$ nanoparticles wrapped by graphene. Advanced Materials. 2012;**24**:1084-1088

[16] Wang F, Zhang K. Reduced graphene oxide-TiO$_2$ nanocomposite with high photocatalystic activity for the degradation of rhodamine B. Journal of Molecular Catalysis A: Chemical. 2011;**345**:101-107

[17] Liu X, Pan L, Lv T, Zhu G, Lu T, Sun Z, et al. Microwave-assisted synthesis of TiO$_2$-reduced graphene

oxide composites for the photocatalytic reduction of Cr (VI). RSC Advances. 2011;**1**:1245-1249

[18] Seema H, Shirinfar B, Shi G, Youn IS, Ahmed N. Facile synthesis of a selective biomolecule chemosensor and fabrication of its highly fluorescent graphene complex. The Journal of Physical Chemistry B. 2017;**121**:5007-5016

[19] Liu C, Teng Y, Liu R, Luo S, Tang Y, Chen L, et al. Fabrication of graphene films on TiO_2 nanotube arrays for photocatalytic application. Carbon. 2011;**49**:5312-5320

[20] Li N, Liu G, Zhen C, Li F, Zhang L, Cheng HM. Battery performance and photocatalytic activity of mesoporous anatase TiO_2 nanospheres/graphene composites by template-free self-assembly. Advanced Functional Materials. 2011;**21**:1717-1722

[21] Du J, Lai X, Yang N, Zhai J, Kisailus D, Su F, et al. Hierarchically ordered macro-mesoporous TiO_2-graphene composite films: Improved mass transfer, reduced charge recombination, and their enhanced photocatalytic activities. ACS Nano. 2010;**5**:590-596

[22] Jiang B, Tian C, Zhou W, Wang J, Xie Y, Pan Q, et al. In situ growth of TiO_2 in interlayers of expanded graphite for the fabrication of TiO_2-graphene with enhanced photocatalytic activity. Chemistry–A European Journal. 2011;**17**:8379-8387

[23] Shirinfar B, Seema H, Ahmed N. Charged probes: Turn-on selective fluorescence for RNA. Organic & Biomolecular Chemistry. 2018;**16**:164-168

[24] Liang Y, Wang H, Casalongue HS, Chen Z, Dai H. TiO_2 nanocrystals grown on graphene as advanced photocatalytic hybrid materials. Nano Research. 2010;**3**:701-705

[25] Akhavan O, Ghaderi E. Photocatalytic reduction of graphene oxide nanosheets on TiO_2 thin film for photoinactivation of bacteria in solar light irradiation. The Journal of Physical Chemistry C. 2009;**113**:20214-20220

[26] Jiang B, Tian C, Pan Q, Jiang Z, Wang JQ, Yan W, et al. Enhanced photocatalytic activity and electron transfer mechanisms of graphene/TiO_2 with exposed {001} facets. The Journal of Physical Chemistry C. 2011;**115**:23718-23725

[27] Sun L, Zhao Z, Zhou Y, Liu L. Anatase TiO_2 nanocrystals with exposed {001} facets on graphene sheets via molecular grafting for enhanced photocatalytic activity. Nanoscale. 2012;**4**:613-620

[28] Guo J, Zhu S, Chen Z, Li Y, Yu Z, Liu Q, et al. Sonochemical synthesis of TiO_2 nanoparticles on graphene for use as photocatalyst. Ultrasonics Sonochemistry. 2011;**18**:1082-1090

[29] Perera SD, Mariano RG, Vu K, Nour N, Seitz O, Chabal Y, et al. Hydrothermal synthesis of graphene-TiO_2 nanotube composites with enhanced photocatalytic activity. ACS Catalysis. 2012;**2**:949-956

[30] Liu B, Huang Y, Wen Y, Du L, Zeng W, Shi Y, et al. Highly dispersive {001} facets-exposed nanocrystalline TiO_2 on high quality graphene as a high performance photocatalyst. Journal of Materials Chemistry. 2012;**22**:7484-7491

[31] Zhao D, Sheng G, Chen C, Wang X. Enhanced photocatalytic degradation of methylene blue under visible irradiation on graphene@ TiO_2 dyade structure. Applied Catalysis B: Environmental. 2012;**111**:303-308

[32] Gopalakrishnan K, Joshi HM, Kumar P, Panchakarla LS, Rao CN. Selectivity in the photocatalytic properties of the composites of TiO_2

nanoparticles with B-and N-doped graphenes. Chemical Physics Letters. 2011;**511**:304-308

[33] Seema H, Kemp KC, Le NH, Park SW, Chandra V, Lee JW, et al. Highly selective CO_2 capture by S-doped microporous carbon materials. Carbon. 2014;**66**:320-326

[34] Kamegawa T, Yamahana D, Yamashita H. Graphene coating of TiO_2 nanoparticles loaded on mesoporous silica for enhancement of photocatalytic activity. The Journal of Physical Chemistry C. 2010;**114**:15049-15053

[35] Wen Y, Ding H, Shan Y. Preparation and visible light photocatalytic activity of Ag/TiO_2/graphene nanocomposite. Nanoscale. 2011;**3**:4411-4417

[36] Zhuo S, Shao M, Lee ST. Upconversion and downconversion fluorescent graphene quantum dots: Ultrasonic preparation and photocatalysis. ACS Nano. 2012;**6**:1059-1064

[37] Kim IY, Lee JM, Kim TW, Kim HN, Kim HI, Choi W, et al. A strong electronic coupling between graphene nanosheets and layered titanate nanoplates: A soft-chemical route to highly porous nanocomposites with improved photocatalytic activity. Small. 2012;**8**:1038-1048

[38] Ren L, Qi X, Liu Y, Huang Z, Wei X, Li J, et al. Upconversion-P25-graphene composite as an advanced sunlight driven photocatalytic hybrid material. Journal of Materials Chemistry. 2012;**22**:11765-11771

[39] Yang Y, Ren L, Zhang C, Huang S, Liu T. Facile fabrication of functionalized graphene sheets (FGS)/ZnO nanocomposites with photocatalytic property. ACS Applied Materials & Interfaces. 2011;**3**:2779-2785

[40] Xu T, Zhang L, Cheng H, Zhu Y. Significantly enhanced photocatalytic performance of ZnO via graphene hybridization and the mechanism study. Applied Catalysis B: Environmental. 2011;**101**:382-387

[41] Lv T, Pan L, Liu X, Lu T, Zhu G, Sun Z. Enhanced photocatalytic degradation of methylene blue by ZnO-reduced graphene oxide composite synthesized via microwave-assisted reaction. Journal of Alloys and Compounds. 2011;**509**:10086-10091

[42] Li B, Cao H. ZnO@ graphene composite with enhanced performance for the removal of dye from water. Journal of Materials Chemistry. 2011;**21**:3346-3349

[43] Luo QP, Yu XY, Lei BX, Chen HY, Kuang DB, Su CY. Reduced graphene oxide-hierarchical ZnO hollow sphere composites with enhanced photocurrent and photocatalytic activity. The Journal of Physical Chemistry C. 2012;**116**:8111-8117

[44] Liu S, Tian J, Wang L, Luo Y, Sun X. One-pot synthesis of CuO nanoflower-decorated reduced graphene oxide and its application to photocatalytic degradation of dyes. Catalysis Science & Technology. 2012;**2**:339-344

[45] Seema H, Kemp KC, Chandra V, Kim KS. Graphene-SnO_2 composites for highly efficient photocatalytic degradation of methylene blue under sunlight. Nanotechnology. 2012;**23**:355705

[46] An X, Jimmy CY, Wang Y, Hu Y, Yu X, Zhang G. WO_3 nanorods/graphene nanocomposites for high-efficiency visible-light-driven photocatalysis and NO_2 gas sensing. Journal of Materials Chemistry. 2012;**22**:8525-8531

[47] Liu X, Pan L, Lv T, Zhu G, Sun Z, Sun C. Microwave-assisted synthesis of CdS-reduced graphene oxide composites for photocatalytic reduction

of Cr (vi). Chemical Communications. 2011;**47**:11984-11986

[48] Wang X, Tian H, Yang Y, Wang H, Wang S, Zheng W, et al. Reduced graphene oxide/CdS for efficiently photocatalytic degradation of methylene blue. Journal of Alloys and Compounds. 2012;**524**:5-12

[49] Ye A, Fan W, Zhang Q, Deng W, Wang Y. CdS–graphene and CdS–CNT nanocomposites as visible-light photocatalysts for hydrogen evolution and organic dye degradation. Catalysis Science & Technology. 2012;**2**:969-978

[50] Hu H, Wang X, Liu F, Wang J, Xu C. Rapid microwave-assisted synthesis of graphene nanosheets–zinc sulfide nanocomposites: Optical and photocatalytic properties. Synthetic Metals. 2011;**161**:404-410

[51] Xiong Z, Zhang LL, Zhao XS. Visible-light-induced dye degradation over copper-modified reduced graphene oxide. Chemistry–A European Journal. 2011;**17**:2428-2434

[52] Xiong Z, Zhang LL, Ma J, Zhao XS. Photocatalytic degradation of dyes over graphene-gold nanocomposites under visible light irradiation. Chemical Communications. 2010;**46**:6099-6101

[53] Fu Y, Wang X. Magnetically separable ZnFe$_2$O$_4$-graphene catalyst and its high photocatalytic performance under visible light irradiation. Industrial & Engineering Chemistry Research. 2011;**50**:7210-7218

[54] Fu Y, Xiong P, Chen H, Sun X, Wang X. High photocatalytic activity of magnetically separable manganese ferrite-graphene heteroarchitectures. Industrial & Engineering Chemistry Research. 2012;**51**:725-731

[55] Fu Y, Chen H, Sun X, Wang X. Graphene-supported nickel ferrite: A magnetically separable photocatalyst

with high activity under visible light. AIChE Journal. 2012;**58**:3298-3305

[56] Min YL, Zhang K, Chen YC, Zhang YG. Enhanced photocatalytic performance of Bi$_2$WO$_6$ by graphene supporter as charge transfer channel. Separation and Purification Technology. 2012;**86**:98-105

[57] Zhou F, Shi R, Zhu Y. Significant enhancement of the visible photocatalytic degradation performances of γ-Bi$_2$MoO$_6$ nanoplate by graphene hybridization. Journal of Molecular Catalysis A: Chemical. 2011;**340**:77-82

[58] Zhang X, Quan X, Chen S, Yu H. Constructing graphene/InNbO$_4$ composite with excellent adsorptivity and charge separation performance for enhanced visible-light-driven photocatalytic ability. Applied Catalysis B: Environmental. 2011;**105**:237-422

[59] Ying H, Wang ZY, Guo ZD, Shi ZJ, YANG SF. Reduced graphene oxide-modified Bi$_2$WO$_6$ as an improved photocatalyst under visible light. Acta Physico-Chimica Sinica. 2011;**27**:1482-1486

[60] Bai S, Shen X, Zhong X, Liu Y, Zhu G, Xu X, et al. One-pot solvothermal preparation of magnetic reduced graphene oxide-ferrite hybrids for organic dye removal. Carbon. 2012;**50**:2337-2346

[61] Gao E, Wang W, Shang M, Xu J. Synthesis and enhanced photocatalytic performance of graphene-Bi$_2$WO$_6$ composite. Physical Chemistry Chemical Physics. 2011;**13**:2887-2893

[62] Zhang H, Fan X, Quan X, Chen S, Yu H. Graphene sheets grafted Ag@AgCl hybrid with enhanced plasmonic photocatalytic activity under visible light. Environmental Science & Technology. 2011;**45**:5731-5736

[63] Chen P, Xiao TY, Li HH, Yang JJ, Wang Z, Yao HB, et al. Nitrogen-doped graphene/ZnSe nanocomposites: Hydrothermal synthesis and their enhanced electrochemical and photocatalytic activities. ACS Nano. 2011;**6**:712-719

[64] Zhang LL, Xiong Z, Zhao XS. Pillaring chemically exfoliated graphene oxide with carbon nanotubes for photocatalytic degradation of dyes under visible light irradiation. ACS Nano. 2010;**4**:7030-7036

[65] Krishnamoorthy K, Mohan R, Kim SJ. Graphene oxide as a photocatalytic material. Applied Physics Letters. 2011;**98**:244101

[66] Zhang Q, He Y, Chen X, Hu D, Li L, Yin T, et al. Structure and photocatalytic properties of TiO_2-graphene oxide intercalated composite. Chinese Science Bulletin. 2011;**56**:331-339

[67] Yoo DH, Cuong TV, Pham VH, Chung JS, Khoa NT, Kim EJ, et al. Enhanced photocatalytic activity of graphene oxide decorated on TiO_2 films under UV and visible irradiation. Current Applied Physics. 2011;**11**:805-808

[68] Nguyen-Phan TD, Pham VH, Shin EW, Pham HD, Kim S, Chung JS, et al. The role of graphene oxide content on the adsorption-enhanced photocatalysis of titanium dioxide/graphene oxide composites. Chemical Engineering Journal. 2011;**170**:226-232

[69] Liu J, Bai H, Wang Y, Liu Z, Zhang X, Sun DD. Self-assembling TiO_2 nanorods on large graphene oxide sheets at a two-phase interface and their anti-recombination in photocatalytic applications. Advanced Functional Materials. 2010;**20**:4175-4181

[70] Liu J, Liu L, Bai H, Wang Y, Sun DD. Gram-scale production of graphene oxide-TiO_2 nanorod composites: Towards high-activity photocatalytic materials. Applied Catalysis B: Environmental. 2011;**106**:76-82

[71] Jiang G, Lin Z, Chen C, Zhu L, Chang Q, Wang N, et al. TiO_2 nanoparticles assembled on graphene oxide nanosheets with high photocatalytic activity for removal of pollutants. Carbon. 2011;**49**:2693-2701

[72] Chen C, Cai W, Long M, Zhou B, Wu Y, Wu D, et al. Synthesis of visible-light responsive graphene oxide/TiO_2 composites with p/n heterojunction. ACS Nano. 2010;**4**:6425-6432

[73] Liao G, Chen S, Quan X, Yu H, Zhao H. Graphene oxide modified gC_3N_4 hybrid with enhanced photocatalytic capability under visible light irradiation. Journal of Materials Chemistry. 2012;**22**:2721-2726

[74] Zhu M, Chen P, Liu M. Graphene oxide enwrapped Ag/AgX (X = Br, Cl) nanocomposite as a highly efficient visible-light plasmonic photocatalyst. ACS Nano. 2011;**5**:4529-4536

Chapter 3

Enhancing Liquid Phase Exfoliation of Graphene in Organic Solvents with Additives

Randhir Singh Bhoria

Abstract

Graphene is the wonder carbon nanomaterial with excellent electrical, mechanical, chemical and optical properties suitable for the fabrication of modern electronics devices such as supercapacitors, sensors, FET etc. Liquid phase exfoliation is the economical, safe, facile method of graphene synthesis without the requirement of harmful chemicals, toxic gases. However, the low concentration of graphene (<0.01 mg/ml) obtained by this method limits its application in various fields. Various techniques have been employed for enhancing the graphene concentration in certain organic solvents. Addition of additives and salts can enhance the graphene concentration in organic solvents to some extent. In this chapter, the earlier work done in enhancing graphene concentration is explained. Further, this technique is employed for graphene concentration enhancement in solvents by using new salts and additives. The results obtained with various additives are compared and it was found that by adding anthracene in NMP solvent graphene concentration increases upto 0.04 mg/ml. This process can be easily scaled up for better performance, and resulting high concentration graphene can be used for the fabrication of the efficient modern electronics devices.

Keywords: graphene synthesis, liquid phase exfoliation, additives, concentration enhancement, organic solvents

1. Introduction

Graphene is the most studied and explored nanomaterial with exceptional mechanical, electrical, optical and chemical properties. Graphene was discovered by Kostya Novoselov and Andre Geim via mechanical exfoliation method. Graphene has 2D hexagonal honeycomb structure made up of carbon atoms. Graphene is highly transparent as it reflects 2.3% and transmits 97.7% of light falling on it which makes it highly useful for making transparent conducting electrodes. Other exotic properties of graphene are high carrier mobility (200,000 $cm^2\,v^{-1}\,s^{-1}$), Young's modulus of 1.0 TPa. Graphene is approximately 200 times more conductive than copper and 100 times stronger than steel. In addition to this it is very flexible in nature as it can be stretched to 20% of its original length.

Because of its very high electrical conductivity, high transparency and flexibility it is being used for the fabrication of wide variety of devices such as flexible transparent displays, energy storage devices etc. Graphene can be prepared by various methods such as mechanical exfoliation, electrochemical, liquid phase exfoliation and CVD method. This chapter uses liquid phase exfoliation method for the synthesis and concentration enhancement of graphene, after discussing its relative advantages over other techniques.

2. Graphene synthesis methods

After the discovery of graphene a lot of research has been done for finding a suitable technique of graphene synthesis. Graphene synthesis method should be facile, economical and can be performed easily in the laboratory and should not require sophisticated equipment. Some of the most prominent graphene synthesis methods are described here (**Figure 1**).

2.1 Mechanical exfoliation method

Mechanical exfoliation method is the oldest and the simplest method, can be easily used in the college lab to prepare graphene of few micrometer length. It is called scotch tape method because here, a scotch tape repeatedly peels off various layers of graphene from the graphite source [1]. The disadvantage of this method is that it is time consuming and does not give graphene sheets of uniform thickness. Moreover, it is not a scalable method to produce high quality graphene sheets.

2.2 Chemical exfoliation method

Chemical exfoliation method uses harmful acids for the exfoliation of graphene sheets from graphite source. When graphite powder is mixed in solution of sulfuric acid and nitric acid, the inter-planar distance between individual graphene sheets

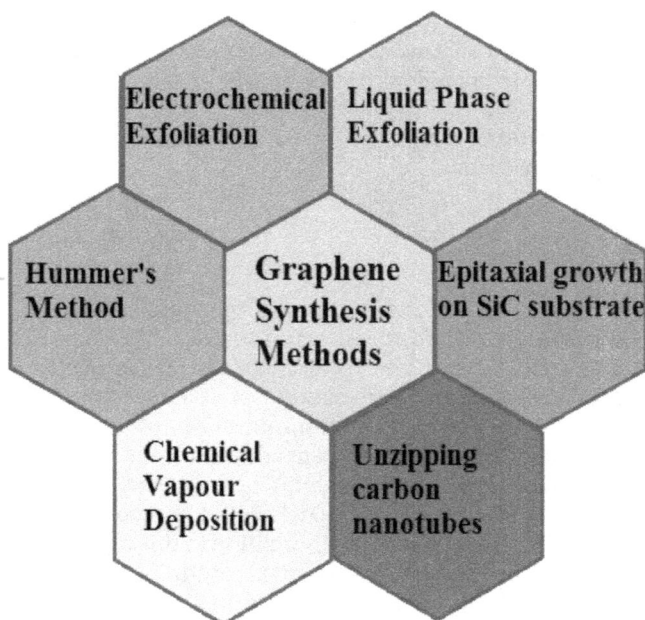

Figure 1.
Various synthesis methods of graphene.

increases and hence exfoliation occurs [2]. The technique of increasing the inter-planar distance between graphene sheets is called as intercalation [3]. The advantage of the intercalation is that after intercalation, the intercalated graphite can be easily exfoliated via sonication. Other processes such as ultrasonic heating [4], acid treatment [5] are used for synthesis of graphene nanoribbons.

2.3 Chemical vapor deposition (CVD) method

CVD technique is used to produce good quality graphene on various substrates such as copper, nickel, cobalt etc. Here, the substrate is placed inside a furnace and hydrocarbon gas is passed at high temperatures. The carbon present in the hydrocarbon gas gets deposited on the substrate to form a graphene layer. Usually a mixture of hydrogen, argon and methane gas is passed through the furnace at temperature of 750–1200°C (**Figure 2**).

The CVD method is useful for scalable synthesis of high quality graphene. Using this method graphene has been successfully deposited upon various substrates such as Ni [6], Rh [7], Pt [8–11], Ir [12], Ru [13–16], Pd [17], Cu [18–21] using methane and ethylene. Graphene has also been prepared by using table sugar as a solid carbon source [22].

2.4 Epitaxial growth on SiC

Graphene layer can be grown on the silicon carbide substrate by heating at high temperature greater than 1100°C (**Figure 3**). The thickness of the graphene film which is prepared on SiC substrate depends upon size of the silicon carbide substrate because Si atoms are desorbed from surface during this process [23–25].

Whether SiC face is silicon or carbon terminated, graphene layer thickness and carrier density changes accordingly [26].

2.5 Hummer's method

Hummer's method is a well-known method to prepare high yield graphene. Here, graphite powder is first converted to graphite oxide with potassium permanganate, which is further converted into graphene oxide (GO) via sonication process. Graphene oxide is later reduced to reduced graphene oxide (rGO) using various reducing agents such as hydrazine hydrate (**Figure 4**).

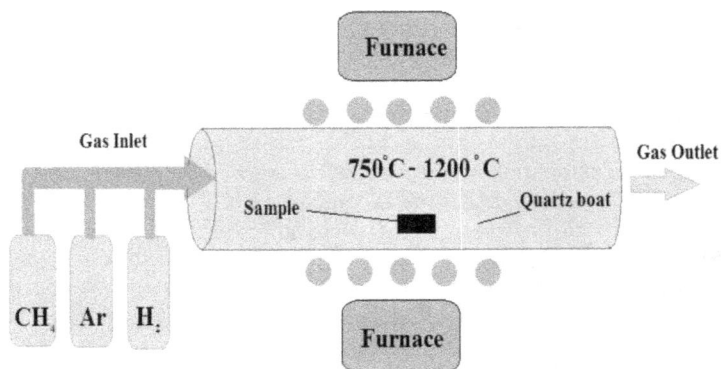

Figure 2.
Schematic illustration of CVD method of graphene synthesis.

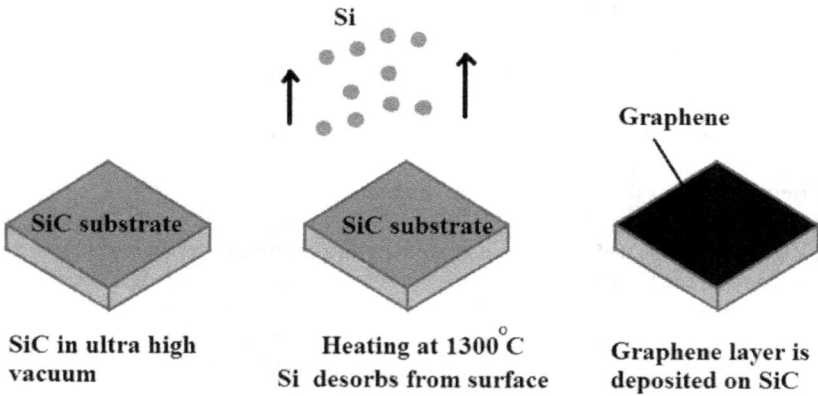

Figure 3.
Process of epitaxial growth of graphene on SiC surface.

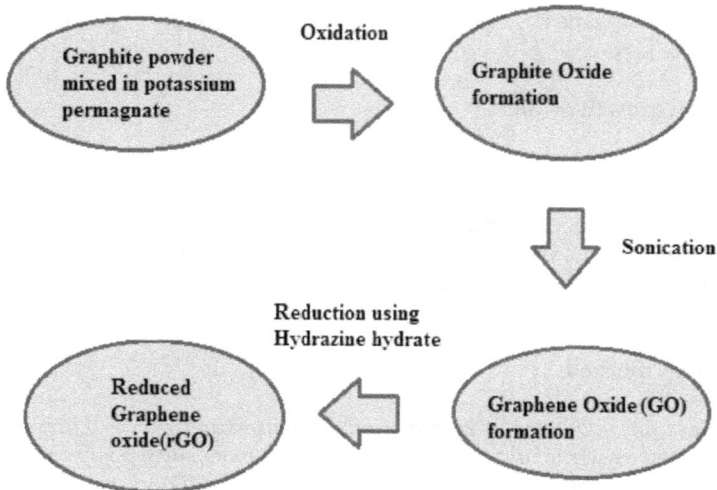

Figure 4.
Synthesis of reduced graphene oxide using Hummer's method.

Most of rGO properties match with graphene, but due to structural defects in rGO, it does not produce high quality graphene. Moreover, because of the use of harmful and toxic chemicals, it is also not safe method.

2.6 Electrochemical exfoliation method

Electrochemical exfoliation method is another method for producing graphene from graphite rod in much shorter time as compared to CVD and Hummer's methods.

Typically, a platinum wire acts as cathode and graphite rod acts as anode (**Figure 5**).

Both anode and cathode are dipped in an electrolyte solution which is usually an acid solution such as sulfuric acid, phosphoric acid etc. After applying 10 V DC between anode and cathode, graphene exfoliation starts [27]. After 2 h graphene nanosheets accumulated in electrolyte are filtered, washed and dried for characterization using SEM, TEM etc. In this method of graphene synthesis also uses various toxic and harmful acids and chemicals.

2.7 Liquid phase exfoliation

Liquid phase exfoliation of graphene uses sonication process to exfoliate graphene from graphite in solvents. Graphite has various layers of graphene attached by Van der Waals forces which is overcome if the solvent used has surface tension near 40–50 mJ/m^2 range. Some of the commonly used solvents are DMF (N,N-Dimethylformamide) and ODCB (ortho-dichlorobenzene). Typically, 2 g graphite powder is added to 300 ml of ODCB. This mixture is sonicated for 3 h. Then it is centrifuged for 30 min at 4000 rpm. (**Figure 6**) Finally supernatants are used for characterization using SEM and TEM [28].

Table 1 shows that various synthesis methods have been explored by the researchers for graphene synthesis each having some disadvantages [29–34]. Hummer's method usually gives high-yield graphene but suffers from defects and impurities in graphene structure. So, rGO prepared by Hummer's method is not useful for the fabrication of electronics devices [35, 36]. Electronics devices fabrication requires good-quality defect-free graphene which can be easily prepared by

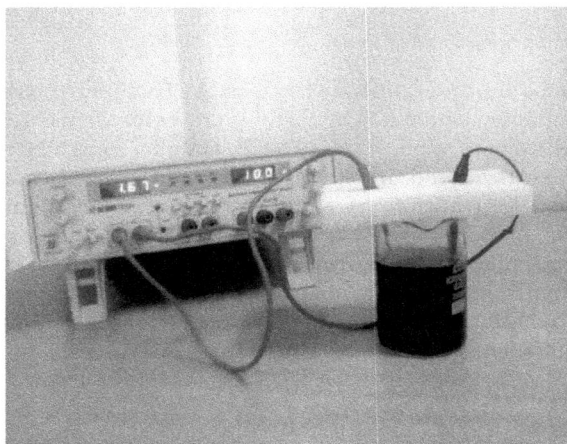

Figure 5.
Electrochemical exfoliation of graphene.

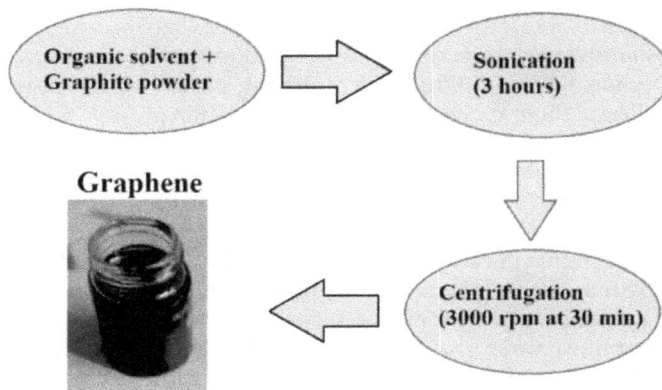

Figure 6.
Various steps involved in liquid phase exfoliation of graphene.

Synthesis method	Advantages	Disadvantages
Mechanical exfoliation method	Good quality, low yield	Not a scalable process, low yield
Chemical vapor deposition	High quality, large area graphene	High temperature and low vacuum conditions
Epitaxial growth on SiC	Large continuous film, good quality	High temperature and low vacuum conditions Not transferable
Hummer's method	High yield	High defects in graphene, harmful chemicals used
Electrochemical exfoliation	Lesser time, facile, economical	DC voltage and electrolytes requirements
Liquid phase exfoliation	*Easy, safe, high quality, economical*	*Long sonication time requirement, low graphene concentration*

Note: The text is in italics to indicate that in this chapter liquid phase exfoliation method is used for the concentration enhancement of graphene.

Table 1.
Various graphene synthesis methods with advantages and disadvantages.

liquid-phase exfoliation, and cannot be produced by Hummer's method [37–41]. The high quality graphene prepared by liquid-phase exfoliation is suitable for modern electronics device applications [42–47]. Graphene has high quantum capacitance [48], electrochemical properties [49] which can help to detect various explosives such as 2,4,6-trinitrotoluene [50], and generate chemiluminescence [51].

Although liquid phase exfoliation technique is the safe and environment friendly technique as compared to other graphene synthesis method, but its disadvantage is lower concentration of graphene obtained (usually less than 0.01 mg/ml). Therefore, there is a need of concentration enhancement of graphene for the fabrication of better and more efficient electronic devices.

3. Process of concentration enhancement

It has been experimentally found that graphene concentration can be enhanced by increasing sonication process time upto many weeks instead of hours which create defects in the graphene nanosheets [52]. Other approaches which have been utilized are mixed solvents [53–55], solvent exchange [56], solvothermal exfoliation [57, 58], intercalants [59–64]. It has been observed that Sodium hydroxide and naphthalene can enhance the graphene concentration in organic solvent [65, 66]. It has been experimentally found that solvents having surface tension near 40–50 mJ/m^2 are highly useful for graphene synthesis via sonication process. Some of the solvents satisfying this criterion such as 1-methyl-2-pyrrolidinone (NMP), benzyl benzoate (BB), 1,2-dichlorobenzene (ODCB), acetophenone (ACP), benzonitrile (BZN), dimethyl sulfoxide (DMSO) and 1,4-dioxane are used in concentration enhancement. It has been found that addition of organic salts can enhance the grapheme concentration in various organic solvents [67]. In addition to various salts, some additives such as phenolphthalein and anthracene have been explored for enhancing graphene concentration in solvents.

Initially, graphite powder is added to NMP, DMSO and CYN solvents (100 ml) with a concentration of 10 mg/ml. Then 100 mg of the additive or salt is added and sonicated for 3 h and centrifuged at 3000 rpm for 30 min (**Figure 7**). After centrifugation, supernatants are used for characterization using UV-Vis spectrum for finding graphene concentration by applying Lambert Beer's law.

UV-Vis spectrum plots the variation of absorbance with wavelength. This technique is based on the principle that the absorption in a particular wavelength range is directly proportional to the color of the sample used for characterization. From the UV-Vis spectrum it is observed that at a particular wavelength the absorbance is maximum. For pure graphene sample the peak absorbance is obtained near 270 nm (**Figure 8**). As the impurities and functional groups are introduced in graphene the absorbance peak is shifted from 270 nm.

After plotting the UV-Vis spectrum curves we calculated the absorbance value at 660 nm wavelength. This absorbance A can be used to find the concentration of the graphene by applying Lambert Beer's law. So, according to the **Lambert Beer's law** the absorbance in terms of concentration is given by:

$$A = \alpha \, C \, l. \tag{1}$$

where **A** is the absorbance measured at **660 nm**,
l is the sample path length which is **1 cm**,

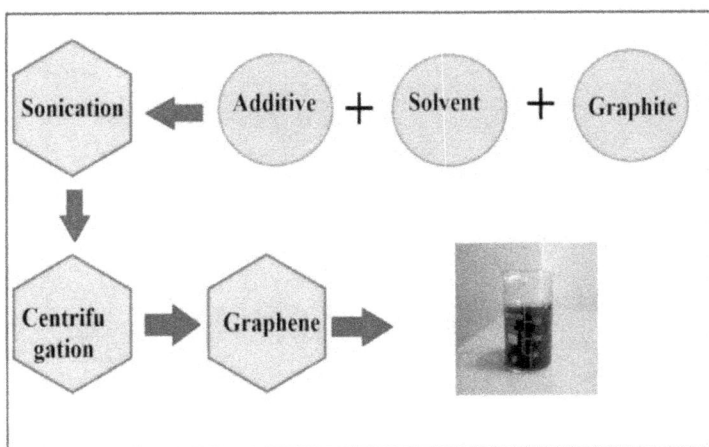

Figure 7.
Schematic diagram of liquid phase exfoliation of graphene with addition of additive.

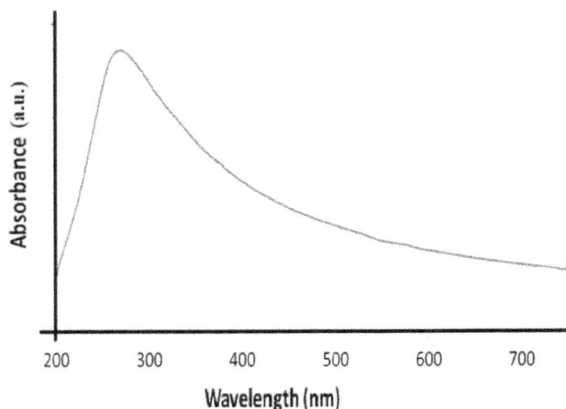

Figure 8.
UV-Vis spectrum of graphene.

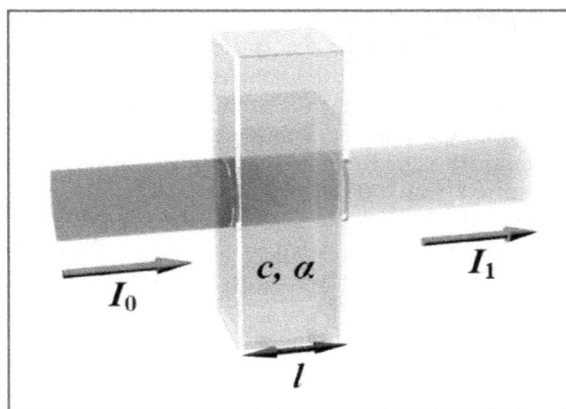

Figure 9.
Representation of Lambert-Beer's law.

Figure 10.
(a) TEM images, (b) SEM images of graphene exfoliated in NMP solvent.

C is the grapheme concentration in the sample,
α is the extinction coefficient.
This process of UV-Vis spectroscopy is shown in **Figure 9** below.

The absorbance value at 660 nm in the UV spectrum of graphene is used to calculate the graphene concentration with absorption coefficient value **α = 2460 ml/mg/m**.

It is observed that additives intercalate onto the graphitic layers which enhances graphene concentration by helping in the exfoliation process. The grapheme nanosheets produced by this process are used for the characterization using SEM and TEM. The TEM results of the grapheme nanosheets shows that single and few-layered, overlapped nanosheets have been produce (**Figure 10**).

Figure 10 shows the SEM results of the graphene nanosheets produced in NMP organic solvent which indicates that graphene sheets size varies from 5 µm to 20 µm.

4. Concentration enhancement with salts and additives

DMSO solvent is used for grapheme concentration enhancement by adding sodium tartrate (ST), sodium chloride (NaCl), potassium chloride (KCl), edetate disodium (ED), sodium citrate (SC), naphthalene (N) and phenolphthalein (P) in it and analyzing their UV spectrum. The absorbance at 660 nm in UV spectrum is used for finding graphene concentration via Lambert Beer's law. By using this law graphene concentration is calculated and plotted in **Figure 11**. From **Figure 11** it is observed that SC and ST produces maximum concentration in DMSO.

The main reason of the concentration enhancement is that additives intercalates onto graphene layers and helps in exfoliation process. It is also observed that lesser time is required for exfoliation after adding additives. Further, in contrast to organic salts, the inorganic salt KCl does not enhances the graphene concentration in DMSO. One reason for low concentration may be the aggregation of graphene nanosheets upon addition of KCl.

Firstly, the UV-Vis spectrum of the various graphene samples (with and without addition of additives in DMSO) is used to find the absorbance value at 660 nm. This absorbance value A, with α the absorption coefficient is used to find graphene concentration as given by Beer's law, $A = \alpha C l$. Here, A is the absorbance at 660 nm, l is sample path length (1 cm), C is concentration and constant $\alpha = 2460$ ml/mg/m. This procedure is repeated to calculate the graphene concentration in other solvents such as NMP, ODCB, BB, ACP, BZN etc. from their UV-Vis spectra.

After repeating this procedure with the CYN solvents it is observed that in CYN maximum concentration is achieved by adding phenolphthalein additive. By adding phenolphthalein additive graphene concentration is increased upto nine times. But, as expected the inorganic salts such as CaCl, KCl does not enhance the graphene concentration. As compared to inorganic salts, organic salts are more useful for concentration enhancement with maximum concentration given by sodium citrate salt.

Figure 12 shows the UV spectra of NMP with and without adding salts and additives. There is a great change in graphene concentration in NMP solvent after adding salts and additives in it with maximum graphene concentration of 0.08 mg/ml in NMP was observed by adding anthracene additive in it. In comparison to organic salts, inorganic salt KCl does not increase the graphene concentration in NMP. This low concentration may be again due to the aggregation effect.

Figure 13 shows the effect of adding three additives anthracene, phenolphthalein, naphthalene in NMP, ODCB, BB, ACP, BZN, DMSO and 1,4-dioxane organic solvents. It has been observed that highest concentration is obtained by adding

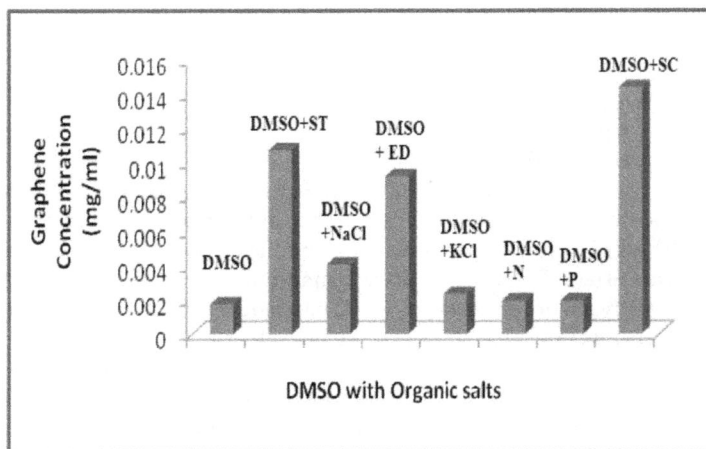

Figure 11.
Concentration variation of graphene in DMSO solvent with addition of salts and additives.

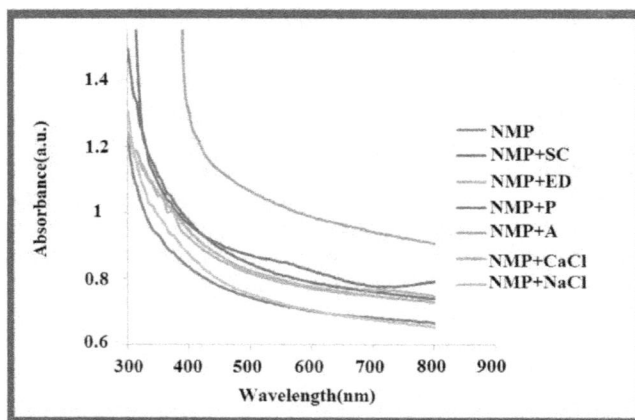

Figure 12.
UV-Vis spectra of graphene in NMP solvent with and without addition of various salts and additives.

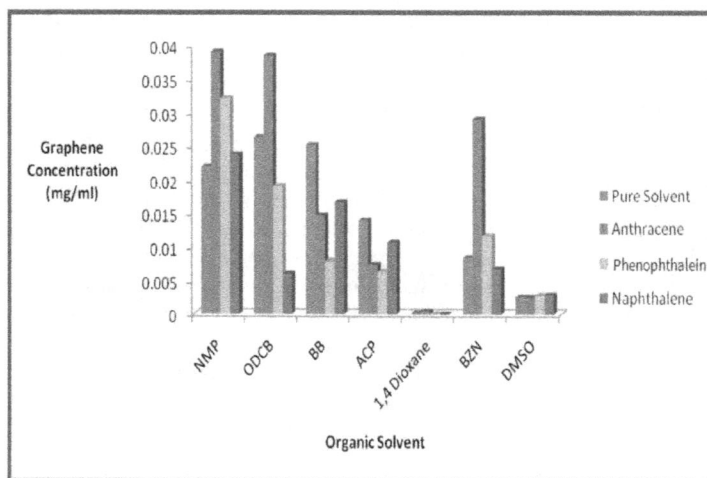

Figure 13.
Graphene concentration variation in seven organic solvents with and without addition of additives.

anthracene in NMP solvent, with 0.04 mg/ml concentration. Anthracene additive acts as molecular wedge between the individual edges of graphite and results in higher concentration. Anthracene when intercalates between the graphene layers, it increases the interplanar spacing between adjacent nanosheets and enhances the exfoliation process. It is also observed that by adding larger amount of anthracene additive (>100 mg) in the solvents, much lower concentration is obtained due to f the aggregation effect. But as the additive amount decreases towards 100 mg, molecules are easily adsorbed onto graphene nanosheets which results in higher graphene concentration.

The structure of the additive and solvents also affects the concentration. Anthracene has structure of three benzene rings. Hence, those solvents which have benzene ring in its structure provides highest graphene concentration enhancement. It is found that solvents such as 1,4-dioxane and DMSO produces minimum graphene concentration as there is an absence of benzene ring structure in them [68]. NMP, BZN and ODCB exhibited maximum concentration of graphene because of benzene ring structures. Two organic solvents BB and ACP have side chains in addition to benzene ring which restricts their intercalation between the individual graphitic layers and hence provides least graphene concentration as compared to other solvents [68].

5. Results and discussions

Various additives and salts have been explored for graphene concentration enhancement in NMP, ODCB, DMF, CYN, DMSO solvents. It has been observed that addition of phenolphthalein in CYN increases graphene concentration from 0.005 to 0.045 mg/ml. With addition of SC salt in DMSO solvent, graphene concentration increased from 0.002 to 0.015 mg/ml which is comparable to earlier published works [27].

The effect of adding Anthracene additive in solvents NMP, DMSO, ODCB, BZN, ACP, BB and 1,4-dioxane have been explored and it has been observed that the concentration depends upon additive-solvent structures and interactions. Because the molecular structure of anthracene has three benzene rings, hence solvents with benzene ring structure produced maximum concentration such as NMP, BZN, ODCB solvents. The results obtained with various additives are compared and it was found that by adding anthracene in NMP solvent graphene concentration increases upto 0.04 mg/ml.

This result is in agreement with earlier published work on naphthalene additive [26]. Here, it is found that anthracene is more useful in BZN solvent (three times concentration) than naphthalene in NMP solvent (two times concentration) [26]. The effect of anthracene in concentration enhancement is most prominent in 1-methyl-2-pyrrolidone (NMP), orthodichlorobenzene (ODCB) and BZN solvents as compared to others. With addition of anthracene, the graphene concentration in NMP and ODCB solvents is increased to 0.04 mg/ml.

From **Figure 13** it is observed that benzyl benzoate (BB) solvent is not very effective in concentration enhancement with anthracene. The least concentration is exhibited by the organic solvents acetophenone, benzyl benzoate, 1,4-dioxane and DMSO. With addition of anthracene, the graphene concentration in NMP and ODCB solvents is increased to 0.04 mg/ml [28]. From **Figure 14** a comparison of

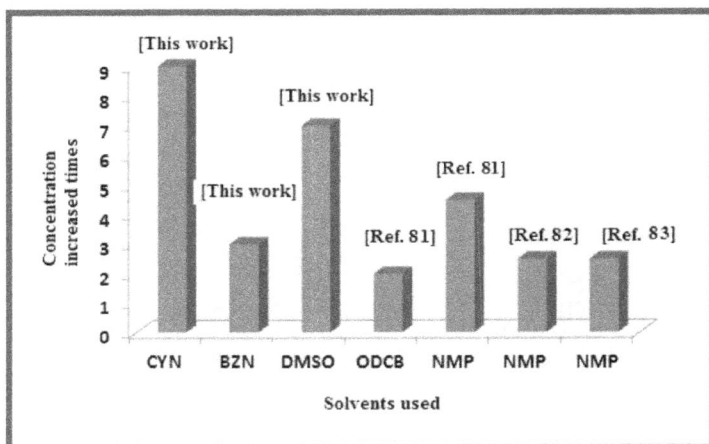

Figure 14.
Comparison of the enhanced concentration in this work with earlier published works.

the graphene concentration enhancement in this work with the earlier published results has been made. The criterion of comparison is the how many times the concentration is increased with additives in various solvents such as CYN, BZN, DMSO, ODCB, NMP etc. From **Figure 14** it is observed that by adding additives graphene concentration is increased to approx. nine times in CYN solvent, approx. three times in BZN solvent, approx. seven times in DMSO solvents after experimental verification. On the other hand by observing earlier published results, graphene concentration was increased from 2.5 to 4.5 only in NMP solvent [69–71].

6. Conclusion

Generally liquid phase exfoliation produces graphene in low concentration (<0.01 mg/ml) which is not suitable for electronics device fabrication. Various additives and salts have been explored for enhancing graphene concentration in organic solvents such as DMSO, CYN, NMP, ODCB, DMF etc. In this study new additives such as phenolphthalein and anthracene have been found to enhance the graphene concentration in solvents such as NMP, ODCB, DMSO and BZN etc. By adding phenolphthalein in cyclohexanone (CYN) solvent the graphene concentration was increased from 0.005 to 0.045 mg/ml and by adding sodium citrate (SC) organic salt, the graphene concentration in dimethyl sulfoxide (DMSO) was increased from 0.002 to 0.015 mg/ml. Further, the effect of addition of anthracene additive in seven organic solvents NMP, DMSO, ODCB, BZN, ACP, BB and 1,4-dioxane have been studied. It was observed that the additive-solvent interactions affect the graphene production yield. Since, the molecular structure of additive anthracene having three benzene rings matches with structure of the solvents such as NMP, BZN, hence causes enhancement in the graphene concentration. The results obtained with various additives are compared and it was found that by adding anthracene in NMP solvent graphene concentration increases upto 0.04 mg/ml. A significant concentration enhancement (three times) was observed with addition of anthracene additive in BZN solvent as compared to two times enhancement with addition of naphthalene in NMP solvent.

Author details

Randhir Singh Bhoria
University Institute of Engineering and Technology, Kurukshetra University, Kurukshetra, India

*Address all correspondence to: mr_randhir_singh@yahoo.co.in

IntechOpen

References

[1] Geim A, Novoselov K. Nature Materials. 2007;**6**:183-191

[2] Hernandez Y, Nicolosi V, Lotya M, Blighe F, Sun Z, De S, et al. Nature Nanotechnology. 2008;**3**:563-568

[3] Viculis L, Mack J, Kaner R. Science. 2003;**299**:1361

[4] Chen G, Weng W, Wu D, Wu C, Lu J, Wang P, et al. Carbon. 2004;**42**:753-759

[5] Li X, Wang X, Zhang L, Lee S, Dai H. Science. 2008;**319**:1229

[6] Mattevi C, Kim H, Chhowalla M. Journal of Materials Chemistry. 2011;**21**:3324

[7] Eizenberg M, Blakely JM. Surface Science. 1970;**82**:228

[8] Castner DG, Sexton BA, Somorjai GA. Surface Science. 1978;**71**:519

[9] Lang B. Surface Science. 1975;**53**:317

[10] Land TA, Michely T, Behm RJ, Hemminger JC, Comsa G. Surface Science. 1992;**264**:261

[11] Sasaki M, Yamada Y, Ogiwara Y, Yagyu S, Yamamoto S. Physical Review B. 2000;**61**:15653

[12] N'Diaye A, Coraux J, Plasa T, Busse C, Michely T. New Journal of Physics. 2008;**10**:043033

[13] Sutter PW, Flege JI, Sutter EA. Nature Materials. 2008;**7**:406

[14] Wu M-C, Xu Q, Goodman DW. The Journal of Physical Chemistry. 1994;**98**:5104

[15] Marchini S, Günther S, Wintterlin J. Physical Review B. 2007;**76**:075429

[16] Vazquez de Parga AL, Calleja F, Borca B, Passeggi MCG Jr, Hinarejos JJ, Guinea F, et al. Physical Review Letters. 2008;**100**:056807

[17] Kwon SY, Ciobanu CV, Petrova V, Shenoy VB, Bareño J, Gambin V, et al. Nano Letters. 2009;**9**:395

[18] Kim KS, Zhao Y, Jang H, Lee SY, Kim JM, Kim KS, et al. Nature. 2009;**457**:706

[19] Jian X et al. Science. 2009;**323**:1701

[20] Reina A, Jia X, Ho J, Nezich D, Son H, Bulovic V, et al. Nano Letters. 2009;**9**:30

[21] Li X et al. Science. 2009;**324**:1312

[22] Sun Z, Yan Z, Yao J, Beitler E, Zhu Y, Tour JM. Nature. 2010;**468**:549

[23] Van Bommel AJ, Crombeen JE, van Tooren A. Surface Science. 1975;**48**:463

[24] Forbeaux I, Themlin J-M, Debever J-M. Physics Review B. 1998;**58**:16396

[25] Charrier A, Coati A, Argunova TJ, Thibaudau F, Garreau Y, Pinchaux R, et al. Journal of Applied Physics. 2002;**92**:2479

[26] Emtsev KV, Seyller T, Ley L, Tadich A, Broekman L, Riley JD, et al. Surface Science. 2006;**600**:3845

[27] Parvez K et al. ACS Nano. 2013;7:3598-3606

[28] Hamilton CE et al. Nano Letters. 2009;**10**:3460-3462

[29] Pham VH, Pham HD, Dang TT, Hur SH, Kim EJ, Kong BS, et al. Journal of Materials Chemistry. 2012;**22**:10530

[30] Park S, Ruoff RS. Nature Nanotechnology. 2009;**4**:217

[31] Park KH, Kim BH, Song SH, Kwon J, Kong BS, Kang K, et al. Nano Letters. 2012;**12**:2871

[32] Lu WB, Liu S, Qin XY, Wang L, Tian JQ, Luo YL, et al. Journal of Materials Chemistry. 2012;**22**:8775

[33] Tien HN, Luan VH, Lee TK, Kong BS, Chung JS, Kim EJ, et al. Chemical Engineering Journal. 2012;**211**:97

[34] Nguyen-Phan TD, Pham VH, Shin EW, Pham HD, Kim S, Chung JS, et al. Chemical Engineering Journal. 2011;**170**:226

[35] Cai M, Thorpe D, Adamson DH, Schniepp HC. Journal of Materials Chemistry. 2012;**22**:24992

[36] Viet HP, Tran VC, Hur SH, Oh E, Kim EJ, Shin EW, et al. Journal of Materials Chemistry. 2011;**21**:3371

[37] Hirsch A, Englert JM, Hauke F. Accounts of Chemical Research. 2013;**46**:87

[38] Loh KP, Bao Q, Ang PK, Yang J. Materials Chemistry. 2010;**20**:2277

[39] Ou EC, Xie YY, Peng C, Song YW, Peng H, Xiong YQ, et al. RSC Advances. 2013;**3**:9490

[40] Barwich S, Khan U, Coleman JN. Journal of Physical Chemistry C. 2013;**117**:19212

[41] Feng L, Liu Y-W, Tang X-Y, Piao Y, Chen S-F, Deng S-L, et al. Chemistry of Materials. 2013;**25**:4487

[42] Lotya M, Hernandez Y, King PJ, Smith RJ, Nicolosi V, Karlsson LS, et al. American Chemical Society. 2009;**131**:3611

[43] De S, Coleman JN. ACS Nano. 2010;**4**:2713

[44] De S, King PJ, Lotya M, OÕNeill A, Doherty EM, Hernandez Y, Duesberg GS, Coleman JN. Small. 2010;**6**:458

[45] Hernandez Y, Nicolosi V, Lotya M, Blighe FM, Sun Z, De S, et al. Nature Nanotechnology. 2008;**3**:563

[46] Du W, Lu J, Sun P, Zhu Y, Jiang X. Chemical Physics Letters. 2013;**198**:568-569

[47] Lin Y, Jin J, Kusmartsevab O, Song M. Journal of Physical Chemistry C. 2013;**117**:17237

[48] Jilin X, Fang C, Jinghong L, Nongjian T. Nature Nanotechnology. 2009;**4**:505-509

[49] Tang L, Wang Y, Li Y, Feng H, Jin L, Li J. Advanced Functional Materials. 2009;**19**:2782-2789

[50] Tang L, Feng H, Cheng J, Li J. Chemical Communications. 2010;**46**:5882

[51] Chen D, Feng H, Li J. Chemical Reviews. 2012;**112**:6027-6053

[52] Coleman JN. Accounts of Chemical Research. 2013;**46**:14

[53] Oyer AJ, Carrillo J-MY, Hire CC, Schniepp HC, Asandei AD, Dobrynin AV, et al. American Chemical Society. 2012;**134**:5018

[54] Yi M, Shen ZG, Ma SL, Zhang XJ. Journal of Nanoparticle Research. 2012;**1003**:14

[55] Yi M, Shen Z, Zhang X, Ma. Journal of Physics D: Applied Physics. 2013;**46**:025301

[56] Li JT, Ye F, Vaziri S, Muhammed M, Lemme MC, Ostling M. Carbon. 2012;**50**:3113

[57] Qian W, Hao R, Hou Y, Tian Y, Shen C, Gao H, et al. Nano Research. 2009;**2**:706

[58] Tang ZH, Zhuang J, Wang X. Langmuir. 2010;**26**:9045

[59] Yang H, Hernandez Y, Schlierf A, Felten A, Eckmann A, Johal S, et al. Carbon. 2013;**53**:357

[60] An X, Simmons T, Shah R, Wolfe C, Lewis KM, Washing-ton M, et al. Nano Letters. 2010;**10**:4295

[61] Liu WW, Wang JN. Chemical Communications. 2011;**47**:6888

[62] Parviz D, Das S, Ahmed HST, Irin F, Bhattacharia S, Green MJ. ACS Nano. 2012;**6**:8857

[63] Xu L, McGraw J-W, Gao F, Grundy M, Ye Z, Gu Z, et al. Journal of Physical Chemistry C. 2013;**117**:10730

[64] Pykal M, Šafarova K, MachalovaŠiškova K, Jureÿcka P, Bourlinos AB, Zboÿril R, et al. Journal of Physical Chemistry C. 2013;**117**:11800

[65] Wei Liu W, Nong Wang J. The royal society of chemistry. Chemical Communications. 2011;**47**:6888

[66] Xu J, Dang DK, Tran VT, Liu X, Chung JS, et al. Journal of Colloid and Interface Science. 2014;**418**:37-42

[67] Du W, Lu J, Sun P, Zhu Y, Jiang X. Chemical Physics Letters. 2013;**568**:198-201

[68] Singh R, Tripathi CC. Enhancing Liquid-Phase Exfoliation of Graphene with Addition of Anthracene in Organic Solvents. Arabian Journal for Science and Engineering. 2017;**42**(6):2417-2424

[69] Haar S, El Gemayel M, Shin Y, Melinte G, Squillaci MA, Ersen O, et al. Enhancing the liquid-phase exfoliation of graphene in organic solvents upon addition of n-octylbenzene. Nature, Scientific Reports. 2015;**5** (Article number: 16684)

[70] Liua WW, Wang JN. Direct exfoliation of graphene in organic solvents with addition of NaOH. Chemical Communications. 2011;**47**:6888-6890

[71] Wencheng D, Lu J, Sun P, Zhu Y, Jiang X. Organic salt-assisted liquid-phase exfoliation of graphite to produce high-quality graphene. Chemical Physics Letters. 2013;**568-569**:198-201

Section 2

Graphene Applications

Chapter 4

Graphene-Based Heterogeneous Electrodes for Energy Storage

Ning Wang, Haixu Wang, Guang Yang, Rong Sun and Ching-Ping Wong

Abstract

As an intriguing two dimensional material, graphene has attracted intense interest due to its high stability, large carrier mobility as well as the excellent conductivity. The addition of graphene into the heterogeneous electrodes has been proved to be an effective method to improve the energy storage performance. In this chapter, the latest graphene based heterogeneous electrodes will be fully reviewed and discussed for energy storage. In detail, the assembly methods, including the ball-milling, hydrothermal, electrospinning, and microwave-assisted approaches will be illustrated. The characterization techniques, including the x-ray diffraction, scanning electron microscopy, transmission electron microscopy, electrochemical impedance spectroscopy, atomic force microscopy, and x-ray photoelectron spectroscopy will also be presented. The mechanisms behind the improved performance will also be fully reviewed and demonstrated. A conclusion and an outlook will be given in the end of this chapter to summarize the recent advances and the future opportunities, respectively.

Keywords: graphene, heterogeneous electrode, energy storage, hydrothermal, EIS, XPS

1. Introduction

In order to overcome the exhaustion of fossil fuels and to address the ever-growing demands for clean, sustainable and high efficient energy supply [1–3], the advanced energy storage techniques, including the supercapacitors, rechargeable batteries (Li-ion battery (LIB), Na-ion battery (SIB)), fuel cells as well as the solar cells have been widely investigated for the commercial use [4–6]. In the advanced energy storage devices, especially for the rechargeable batteries, the electrode materials should have the following features: high energy density, high working voltage, high power density, long cycling stability, high rate capacity as well as the environmental friendly [7–10].

In the rechargeable batteries, e.g. LIBs, the commercial anode material is graphite, whose theoretical-specific capacity is only 372 mA h/g [10], which cannot meet the requirement of the advanced energy storage techniques as described above. In order to overcome the low specific capacity of the graphite anode, amounts of substitute anode materials, e.g. Si (4200 mAh/g) [11], SnO (790 mAh/g) [12, 13], SnSb (825 mAh/g) [14, 15], Sn (993 mAh/g) [16], SbS_3 (947 mAh/g) [17], have been developed for high-capacity rechargeable batteries (**Figure 1**). However, the cycling stability became the most challenging issue for the high-capacity anode materials

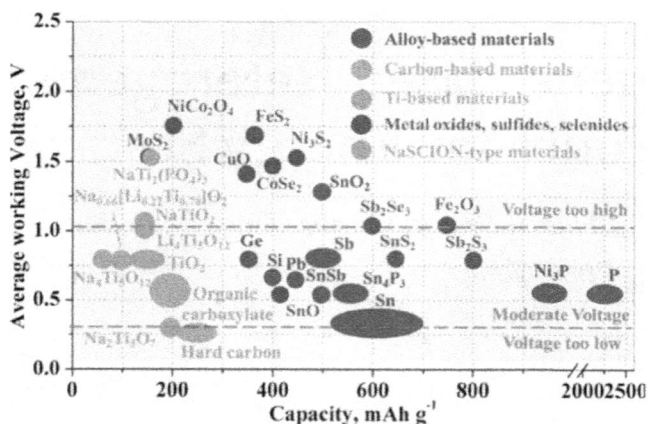

Figure 1.
Performance data of anode materials for SIB, reproduced with permission [19].

due to the volume expansion along with the charge–discharge process [18], e.g. 320% expansion for Si anode. Therefore, the gradient and/or the heterostructured anode materials could be the alternative approaches for the long cycling-stability, high specific capacity rechargeable batteries.

As a promising two dimensional (2D) material, graphene has attracted intense interest in the field of transparent electrode [20–24], field emission transistors (FET) [25–27], flexible devices [28–31], corrosion protection [32–34], catalysis [35–37] and energy storage [38–40], due to its large electrical conductivity, high thermal/chemical stability as well as the flexibility. With respect to the electrode materials, the graphene based heterogeneous electrodes were expected to occupy the excellent electrical conductivity, the long cycling stability and the high rate capability.

In this chapter, the assembly strategies for the graphene based heterogeneous electrodes, including the ball-milling, hydrothermal, electrospinning, microwave-assisted approaches, and the characterization methods will be fully reviewed. The mechanism behind the enhanced performance with graphene will be discussed, and an outlook on the challenges that should be addressed in the future will also be illustrated in the end.

2. Strategies for the assembly

2.1 Ball-milling

As a low-temperature alloying method, ball-milling is highly efficient in preparing the alloys and composites [41–46]. As for the graphene based hetero-geneous electrode materials, ball milling exhibited the advantages in the size/layer reduction [47], interface-contact enhancement [48, 49] as well as the low cost and time saving [50].

As illustrated by Tie et al. [47], in the ball milling preparation of Si@SiO$_x$/graphene heterogeneous anode material (**Figure 2**), the graphene nanosheets (GNS) could be exfoliated from the expanded graphite (EG) due to the accumulated mechanical shearing force of the agate balls, and the particle size of silicon could be reduced to 50–100 nm, which contributed to the uniform dispersion of Si nanoparticles on the GNS, and finally gave rise to the Si@SiO$_x$/graphene composite. Owing to the reduced Si nanoparticle size, the SiO$_x$ adhesion layer as well as the synergistic

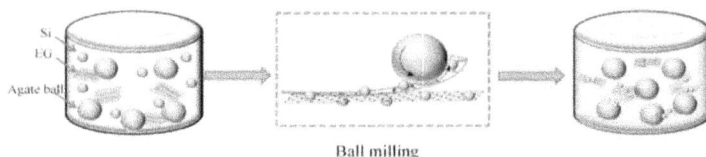

Figure 2.
Schematic illustration for ball milling synthesis of Si@SiO$_x$/grapheme anode material [47].

effect of GNS, the Si@SiO$_x$/graphene heterogeneous anode material exhibited the enhanced cycling stability, high reversible capacity, and rate capability.

Besides, the ball milling method could also be used to prepare other graphene based anode materials. Sun et al. [48] reported the ball milling synthesis of MoS$_2$/graphene anode materials used for high rate SIBs, where the bulky MoS$_2$ and graphite were firstly expanded by the intercalation of Na$^+$ and K$^+$ between the layers, and then the several-layer MoS$_2$ nanosheets and the graphene sheets could be exfoliated from the loose counterparts, which finally resulted in the formation of the restacked MoS$_2$/graphene heterostructures owing to the high surface energy and the interlayer Van der Waals attractions. Chen et al. [51] prepared the center-iodized graphene (CIG) and edge-iodized graphene (EIG) through the ball milling method, and the CIG were found to be an advanced anode material to boost the performance of the LIBs. In the other cases, Xia et al. [52] assembled the layer-by-layered SnS$_2$/graphene anode materials for the LIBs via ball-milling, where the volume change of SnS$_2$ could be buffered by the graphene, and the shuttle effect in the cycling could also be suppressed, both of which gave rise to an excellent rate capability and the negligible capacity fading over 180 cycles; Ma et al. [49] prepared the MoTe$_2$/FLG (few-layer graphene) anode material for the LIBs through the ball milling of MoTe$_2$ and graphite, which exhibited a high reversible capacity and an ultrahigh cycling stability.

2.2 Hydrothermal assembly

Hydrothermal method is an efficient and cost-effective approach for the assembly of metastable crystalline structures [53–57], especially for the heterogeneous structure with solid interface contact [58–61]. As for the graphene based heterogeneous electrode materials, the use of hydrothermal assembly could effectively reduce the cost, improve the crystallinity, and consolidate the interface contact, and therefore improve the energy storage performance.

Pang et al. [62] reported the hydrothermal assembly of VS$_4$@GS (graphene sheets) nanocomposites used as the anode material for the SIBs. As shown in **Figure 3**, the CTA$^+$ (hexadecyl trimethyl ammonium ion) cations were firstly absorbed on the negatively charged GO (graphene oxide) sheets, and then the TAA (thioacetamide) and VO$_4$$^{3-}$ were attached onto the CTA$^+$ to form the TAA-VO$_4$$^{3-}$-CTA$^+$-GO complex, which was then transferred into the VS$_4$/GS composite under the hydrothermal conditions. As an anode material, this composite exhibited a large specific capacity, good rate capability, and remarkable long cycling stability, which should be ascribed to the porous structure together with the synergistic interaction between the highly conductive graphene network and the VS$_4$ nanoparticles.

In other cases, hydrothermal assembly could also be used to fabricate the polyaniline (PANI)/graphene [58, 63], TiO$_2$/graphene [64], Mn$_3$O$_4$/CeO$_2$/graphene [65], α-Fe$_2$O$_3$/graphene [59], and Mn$_3$O$_4$/graphene electrode materials [66]. As illustrated in the literatures, the hydrothermal assembly of graphene based heterogeneous electrode materials is usually starting with the graphite oxide (GO), the active electrode materials and/or surfactants, which should be

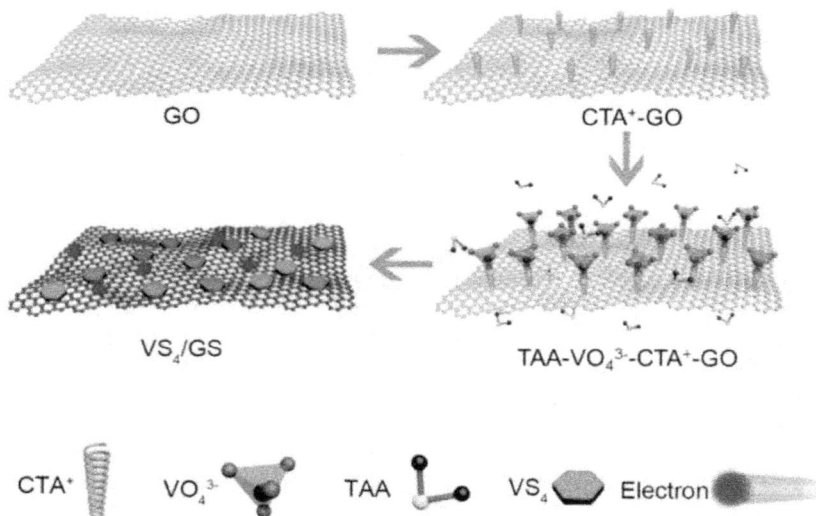

Figure 3.
Hydrothermal synthesis route for the VS$_4$@GS nanocomposites [62].

mainly due to the intrinsic negatively charged surface of the GO that could be easily attached to the positively charged surfactants, and facilitate the nucleation and the growth of active materials on the reduced graphite oxide (rGO, graphene) sheets under the hydrothermal conditions. The strong interface adhesion and the high crystallinity of the hydrothermal assembled composite should benefit the electrode with improved energy storage performance.

2.3 Electrospinning

As an efficient fabrication method for nanofibers [67–73], the electrospinning method has also been developed for producing nanofiber/graphene heterogeneous electrode materials for the energy storage applications [74–78].

As an example, Wei et al. [78] demonstrated an electrospinning fabrication of GO-PAN/PVDF (GPP) membrane electrode for fuel cell applications. In the preparation of GPP membrane electrode material (**Figure 4**), the uniform GPP precursor was prepared by dispersing the PAN, PVDF, and GO in DMF solvent, and then the GPP nanofibers were coated onto the carbon paper sheet attached on a collector drum via the electrospinning. Finally, the electrode was assembled by loading the Pt/C catalysts on the GPP nanofiber membranes.

As a promising procedure, the electrospinning method was also reported to prepare the carbon nanofibers [74], carbonized gold (Au)/graphene (G) hybrid nanowires [75], GO/PVA composite nanofibers [76], and graphene/carbon nanofibers [77] electrode materials for the supercapacitor, biosensor applications. It should be noticed that the uniformity and the viscosity of the precursor should be carefully controlled, since both of which are critical for the mechanical strength and the electrochemical performance of the ultimate products.

2.4 Microwave-assisted assembly

As a quick and even heating method throughout the sample, the microwave assisted heating method has been widely used in the preparation of nanomaterials [79, 80].

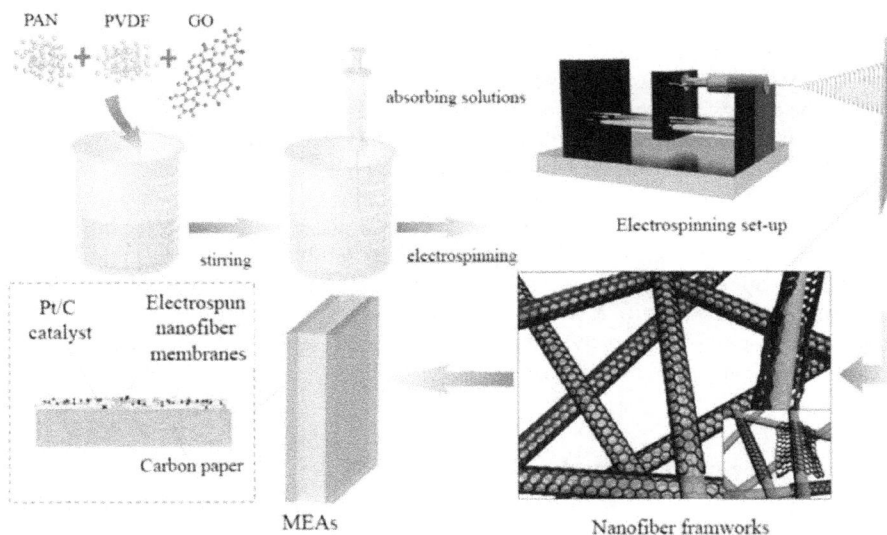

Figure 4.
A synthetic route to GO-PAN/PVDF (GPP) nanofibers [78]. PAN is polyacrylonitrile, and PVDF is polyvinylidene fluoride.

In the preparation of graphene based electrode materials, the microwave assisted method has shown the advantages in the reduction and exfoliation of GO, the time efficiency, and the energy saving [9, 81–83].

As shown in **Figure 5**, Kumar et al. [81] reported the microwave assisted synthesis of palladium (Pd) nanoparticle intercalated nitrogen doped rGO (NrGO) and the application as anode material for the fuel cells. In this synthesis, the GO nanosheets could be reduced and exfoliated under the microwave irradiation with pyridine treatment, and the nitrogen doping could also be achieved via the further

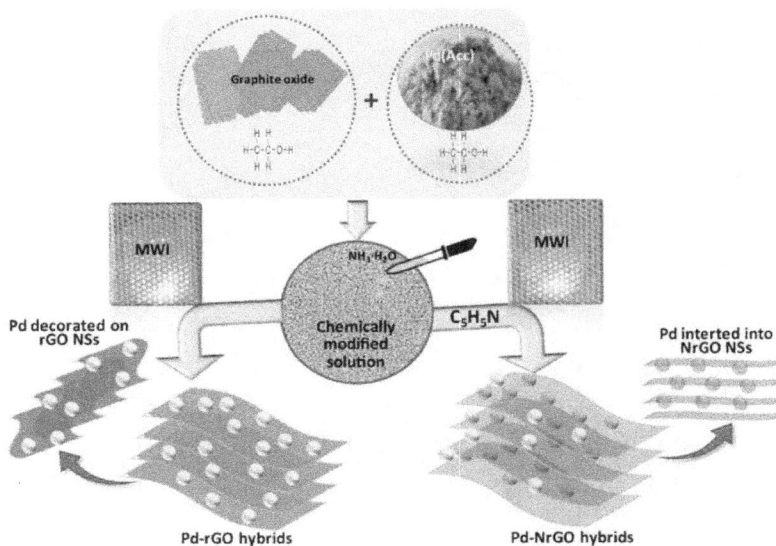

Figure 5.
Schematic illustration of the microwave assisted synthesis of Pd-rGO and Pd-NrGO hybrids [81].

modification with pyridine. The obtained porous rGO and NrGO could be decorated with Pd nanoparticles, which gave rise to a high electroactive surface, and therefore resulted in a high catalytic activity.

For the energy storage electrode materials, the microwave assisted method has been used to ultrafast assembly of the $Mn_{0.8}Co_{0.2}CO_3$/graphene composite [9], SnO_2/graphene composite for LIBs [82], and SnO_2@graphene/N-doped carbons for SIBs [83]. The ultrafast and uniform heating effect of the microwave method should be due to the dielectric heating principle, under which the polar molecules in the microwave radiation could rotate in a high frequency, and thus generate thermal energies evenly across the samples, which benefits the synthesis with environmental friendship, low cost, low energy consumption as well as the porous structures that especially provide the quick transfer channels of the Li^+/Na^+ cations in the rechargeable batteries.

3. Characterization methods

3.1 Scanning electron microscopy (SEM)

In the morphology analysis of the graphene based heterogeneous electrode materials, the top-view and cross-section SEM (**Figure 6**) could be used to determine the distribution of the active materials wrapped or attached by the layered graphene substrates based on the high resolution detector for the secondary electrons emitted on the sample surface. Combined with the EDS (energy dispersive X-ray spectroscopy) technique, the interface of the heterogeneous electrode could also be figured out clearly via the elemental mapping for the active materials and the graphene substrates [48, 51, 82].

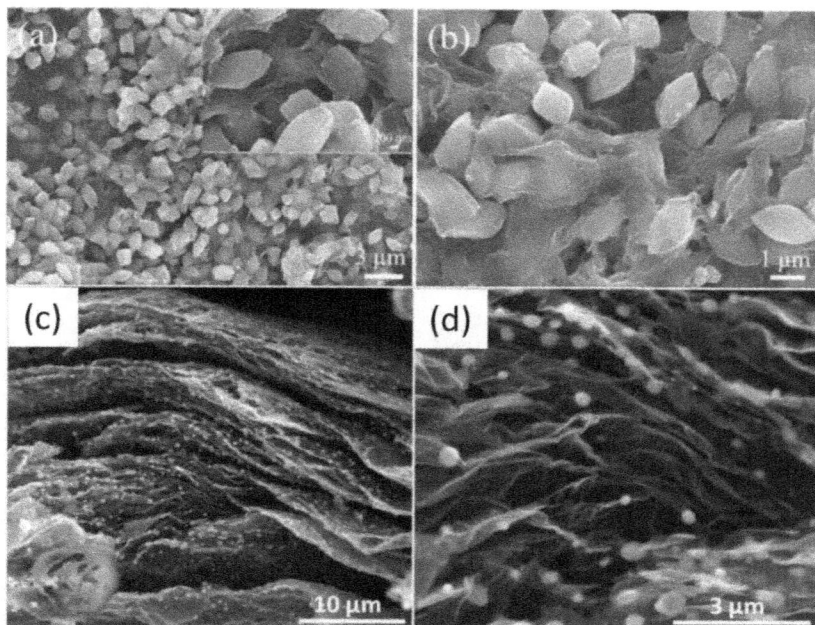

Figure 6.
SEM images for the $Mn_{0.8}Co_{0.2}CO_3$/graphene oxide (a, b) [9] and Pd-NrGO hybrides (c, d) [81].

3.2 Transmission electron microscopy (TEM)

As a powerful characterization method, TEM has been widely used to determine the morphology, crystal structure as well as the interface adhesion of the heterogeneous structures due to its atomic level resolution and the sensitivity to the contrast changes along with the elemental differences on the interface [84, 85]. With respect to the graphene based heterogeneous electrode materials, as shown in **Figure 7**, the uniform dispersion of SnO_2 on the graphene layers could be determined in the low magnification TEM image (**Figure 7b, c**), and the well crystallized SnO_2 nanoparticles could be clearly indexed in the HRTEM (high resolution transmission electron microscopy) and the corresponding FFT (Fast Fourier Transform) patterns (**Figure 7d–i**). The morphology and the crystal structure determined by TEM should be consistent with the result of SEM and XRD, respectively.

3.3 X-ray photoelectron spectroscopy (XPS)

X-ray photoelectron spectroscopy (XPS) is a promising technique for determining the stoichiometry, the valence states, and the bonding conditions of the elements in the compounds, which has been widely used to characterize the functional materials [86–90]. Regarding to the graphene based heterogeneous electrode materials, as shown in **Figure 8** for the high resolution XPS scan of Pd-NrGO hybrids [81], the C 1s XPS peak could be split into the peaks for C=C (284.6 eV), C—O (286.4 eV), C—N

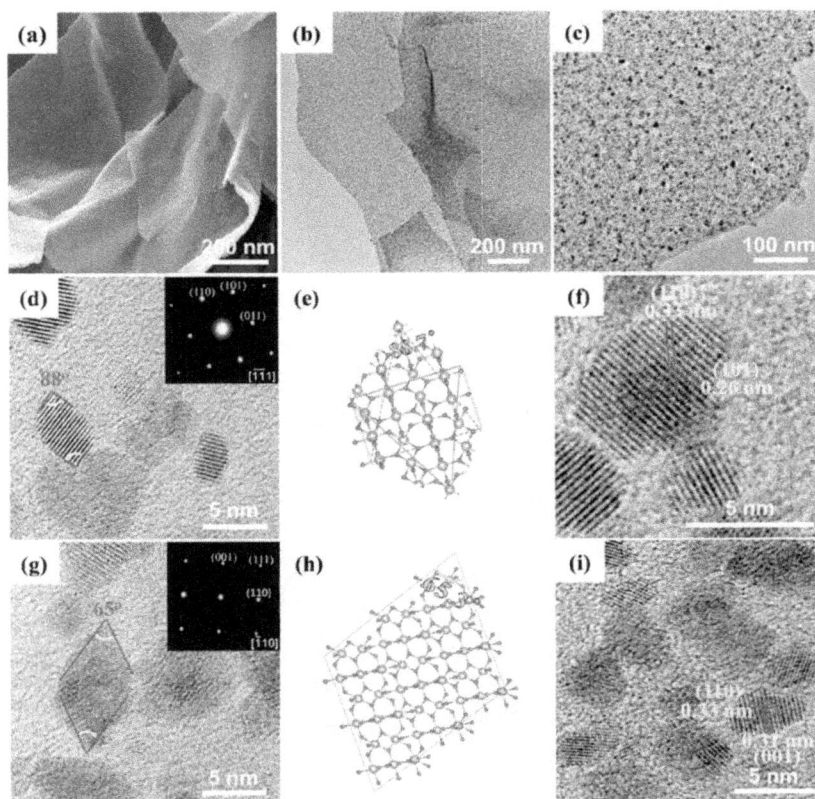

Figure 7.
(a) SEM image. (b, c) Low-magnification TEM images for the SnO_2/graphene hybrids. HRTEM images showing the octahedral SnO_2 model enclosed by {221} facets with (d–f) [$\bar{1}\bar{1}1$] and (g–i) [$\bar{1}01$] zone axes [7].

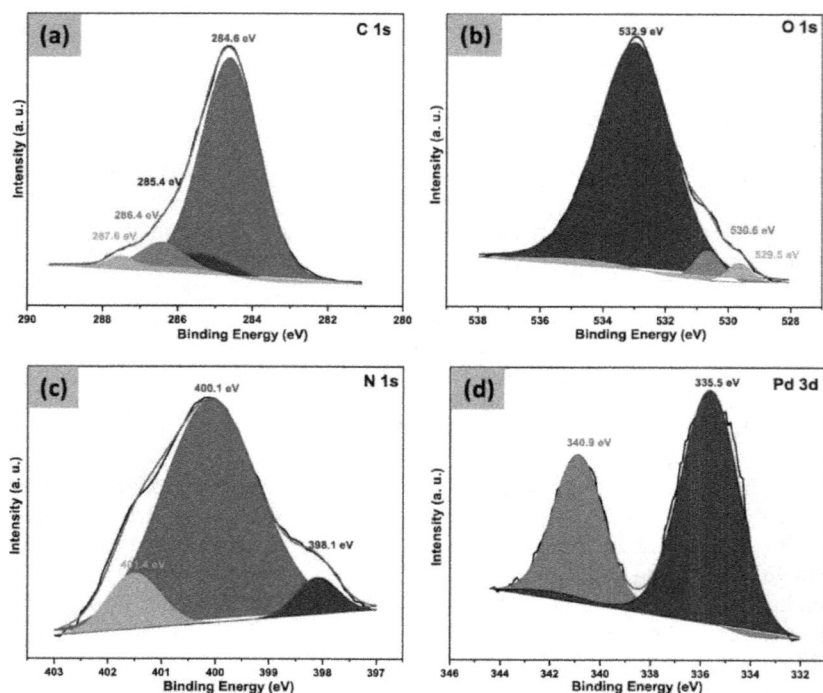

Figure 8.
High resolution XPS spectra for (a) C 1s, (b) O 1s, (c) N 1s and (d) Pd 3d of Pd-NrGO hybrids [81].

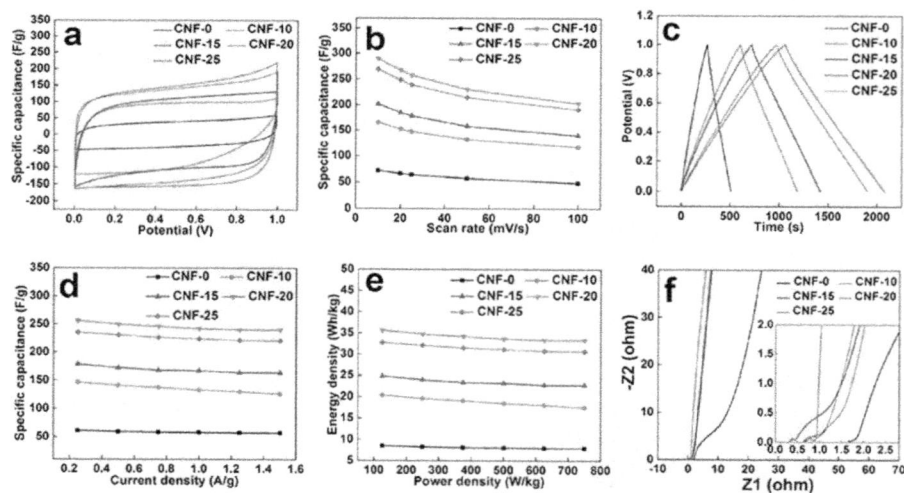

Figure 9.
(a) Cyclic voltammetry (CV) curves of carbon nanofiber samples, (b) rate capability curves of carbon nanofiber from 10 to 100 mV/s, (c) GCD curves of carbon nanofiber samples, (d) rate capability curves of carbon nanofiber from 0.25 to 1.5 a/g, (e) Ragone plots of the supercapacitor devices, and (f) Nyquist plots of carbon nanofiber samples [74].

(285.4 eV), and C=N (287.6 eV), the N 1S peak could be split into the graphitic-N (401.4 eV), pyrollic-N (400.1 eV) and pyridinic-N (398.1 eV) peaks, and the O 1S could be split into the Pd—O (529.5 eV), C—O (530.6 eV), and C=O (532.9 eV) peaks, which fully revealed the bonding information within the Pd-NrGO hybrids.

Apart from the SEM, TEM, and XPS, X-ray diffraction (XRD), Raman, FTIR, and thermal analysis methods (TGA, DSC) were also used to determine the crystal structure, morphology, thermal stability, and other physical/chemical characteristics of the graphene based heterogeneous electrode materials. The electrochemical performance for the energy storage was usually evaluated by the tests, including the cyclic voltammetry (CV), rate capability, galvanostatic charge/discharge (GCD), cycling specific capacity, and the electrochemical impedance spectra (EIS, e.g. Nyquist plots) (**Figure 9**).

4. Mechanisms

As for the active materials in the anode for the energy storage devices (e.g. supercapacitor, LIBs, and SIBs), the modification via bonding or attaching with the graphene or rGO always results in the improvement of the electrochemical performance with respect to the cycling stability, rate capability as well as the high specific capacity.

Behind the enhancement of the performance, there exist several possible mechanisms for the property promotion as illustrated in the following:

a. The growth of nanoparticles for the active materials could be effectively restricted by the graphene, giving rise to the uniform dispersion of the nanoparticles that facilitates the increase of specific area and the active sites for K^+/Na^+ storage [7].

b. The non-faradaic capacitance could be contributed by the graphene due to the electrical double layer-effect [7].

c. The fragmentation of the active materials due to the volume expansion and contraction during the charge–discharge cycles could be depressed by the flexible graphene, which benefits the devices with excellent cycling stability and rate capability [7, 52].

d. The conductivity of the active materials could be enhanced by the graphene, which gives rise to the increase of reversible capacity [52].

e. The graphene in the composite could supply a physical barrier between the active materials and the electrolyte, which effectively suppresses the shuttle effect of the byproducts in the de-charge process that could fade capacity of the batteries [52].

5. Conclusions and outlook

In summary, the synthesis and the characterization of the graphene based heterogeneous electrode materials for the energy storage applications (e.g. SIBs, LIBs, and supercapacitor) have been fully reviewed and discussed in this chapter. In the synthesis of the title materials, ball milling and hydrothermal methods show the cost-effective advantages. Comparatively, the electrospinning method exhibits the benefits in the nanowire composite assembly, and the microwave assisted approach occupies the superiority in the ultrafast fabrication. With respect to the characterization, the morphology could be determined by the SEM and TEM, and the electrochemical performance could be evaluated by the cyclic voltammetry (CV), rate capability, galvanostatic charge/ discharge (GCD), cycling specific capacity, and the EIS tests. In the composite, the graphene could restrict the growth of the nanosized active materials, contribute the non-faradaic capacitance, improve the conductivity, suppress the fragmentation, and supply a physical barrier between the active materials and the electrolyte, which benefit the devices with excellent cycling stability, large rate capability as well as the high specific capacity.

In the future, the most interesting and challenging applications of the graphene in the nanocomposite for the energy storage devices should be the ultrafast rechargeable batteries, the large-energy-density supercapacitors, and the all-solid-state LIBs.

Acknowledgements

This work was supported by the National Key R&D Project from Minister of Science and Technology of China (No. 2017YFB0406200), National and Local Joint Engineering Laboratory of Advanced Electronic Packaging Materials (Shenzhen Development and Reform Committee 2017-934), Leading Scientific Research Project of Chinese Academy of Sciences(QYZDY-SSW-JSC010), and Guangdong Provincial Key Laboratory (2014B030301014).

Conflict of interest

The authors declare no conflict of interest.

Acronyms and abbreviations

LIBs	lithium ion batteries
SIBs	sodium ion batteries
SEM	scanning electron microscopy
TEM	transmission electron microscopy
XRD	X-ray diffraction
XPS	X-ray photoelectron spectroscopy
FTIR	Fourier-transform infrared spectroscopy
TGA	thermal gravimetric analysis
DSC	differential scanning calorimetry

Author details

Ning Wang*, Haixu Wang, Guang Yang, Rong Sun and Ching-Ping Wong
Shenzhen Institutes of Advanced Technology, Chinese Academy of Sciences, Shenzhen, China

*Address all correspondence to: ning.wang@siat.ac.cn

IntechOpen

References

[1] Dunn B, Kamath H, Tarascon J-M. Electrical energy storage for the grid: A battery of choices. Science. 2011;**334**:928-935. DOI: 10.1126/science.1212741

[2] Choi JW, Aurbach D. Promise and reality of post-lithium-ion batteries with high energy densities. Nature Reviews Materials. 2016;**1**:16013. DOI: 10.1038/natrevmats.2016.13

[3] Larcher D, Tarascon JM. Towards greener and more sustainable batteries for electrical energy storage. Nature Chemistry. 2014;**7**:19. DOI: 10.1038/nchem.2085

[4] Slater Michael D, Kim D, Lee E, Johnson Christopher S. Sodium-ion batteries. Advanced Functional Materials. 2012;**23**:947-958. DOI: 10.1002/adfm.201200691

[5] Nayak Prasant K, Yang L, Brehm W, Adelhelm P. From lithium-ion to sodium-ion batteries: Advantages, challenges, and surprises. Angewandte Chemie International Edition. 2017;**57**:102-120. DOI: 10.1002/anie.201703772

[6] Sun Y, Liu N, Cui Y. Promises and challenges of nanomaterials for lithium-based rechargeable batteries. Nature Energy. 2016;**1**:16071. DOI: 10.1038/nenergy.2016.71

[7] Zhang P, Zhao X, Liu Z, Wang F, Huang Y, Li H, Li Y, Wang J, Su Z, Wei G. Exposed high-energy facets in ultradispersed sub-10 nm SnO_2 nanocrystals anchored on graphene for pseudocapacitive sodium storage and high-performance quasi-solid-state sodium-ion capacitors. NPG Asia Materials. 2018;**10**:429-440. DOI: 10.1038/s41427-018-0049-y

[8] Zhang J, Li JY, Wang WP, Zhang XH, Tan XH, Chu WG, Guo YG. Microemulsion assisted assembly of 3D porous S/graphene@ g-C_3N_4 hybrid sponge as free-standing cathodes for high energy density Li-S batteries. Advanced Energy Materials. 2018;**8**:1702839. DOI: 10.1002/aenm.201702839

[9] Xiong Q, Lou J, Zhou Y, Shi S, Ji Z. Ultrafast synthesis of $Mn_{0.8}Co_{0.2}CO_3$/graphene composite as anode material by microwave solvothermal strategy with enhanced Li storage properties. Materials Letters. 2018;**210**:267-270. DOI: 10.1016/j.matlet.2017.09.045

[10] Wang G, Zhang J, Yang S, Wang F, Zhuang X, Müllen K, Feng X. Vertically aligned MoS_2 nanosheets patterned on electrochemically exfoliated graphene for high-performance lithium and sodium storage. Advanced Energy Materials. 2018;**8**:1702254. DOI: 10.1002/aenm.201702254

[11] Chan CK, Peng H, Liu G, McIlwrath K, Zhang XF, Huggins RA, Cui Y. High-performance lithium battery anodes using silicon nanowires. Nature Nanotechnology. 2007;**3**:31. DOI: 10.1038/nnano.2007.411

[12] Aurbach D, Nimberger A, Markovsky B, Levi E, Sominski E, Gedanken A. Nanoparticles of SnO produced by sonochemistry as anode materials for rechargeable lithium batteries. Chemistry of Materials. 2002;**14**:4155-4163. DOI: 10.1021/cm021137m

[13] Kim H, Cho J. Hard templating synthesis of mesoporous and nanowire SnO_2 lithium battery anode materials. Journal of Materials Chemistry. 2008;**18**:771-775. DOI: 10.1039/b714904b

[14] Chao S-C, Song Y-F, Wang C-C, Sheu H-S, Wu H-C, Wu N-L. Study on microstructural deformation of working

Sn and SnSb anode particles for Li-ion batteries by in situ transmission X-ray microscopy. The Journal of Physical Chemistry C. 2011;**115**:22040-22047. DOI: 10.1021/jp206829q

[15] Shi L, Li H, Wang Z, Huang X, Chen L. Nano-SnSb alloy deposited on MCMB as an anode material for lithium ion batteries. Journal of Materials Chemistry. 2001;**11**:1502-1505. DOI: 10.1039/b009907o

[16] Deng D, Kim MG, Lee JY, Cho J. Green energy storage materials: Nanostructured TiO_2 and Sn-based anodes for lithium-ion batteries. Energy & Environmental Science. 2009;**2**:818-837. DOI: 10.1039/b823474d

[17] Xiong X, Wang G, Lin Y, Wang Y, Ou X, Zheng F, Yang C, Wang J-H, Liu M. Enhancing sodium ion battery performance by strongly binding nanostructured Sb_2S_3 on sulfur-doped Graphene sheets. ACS Nano. 2016;**10**:10953-10959. DOI: 10.1021/acsnano.6b05653

[18] Yao Y, McDowell MT, Ryu I, Wu H, Liu N, Hu L, Nix WD, Cui Y. Interconnected silicon hollow nanospheres for Lithium-ion battery anodes with long cycle life. Nano Letters. 2011;**11**:2949-2954. DOI: 10.1021/nl201470j

[19] Xu GL, Amine R, Abouimrane A, Che H, Dahbi M, Ma ZF, Saadoune I, Alami J, Mattis Wenjuan L, Pan F, Chen Z, Amine K. Challenges in developing electrodes, electrolytes, and diagnostics tools to understand and advance sodium-ion batteries. Advanced Energy Materials. 2018;**0**:1702403. DOI: 10.1002/aenm.201702403

[20] Wu J, Becerril HA, Bao Z, Liu Z, Chen Y, Peumans P. Organic solar cells with solution-processed graphene transparent electrodes. Applied Physics Letters. 2008;**92**:263302. DOI: 10.1063/1.2924771

[21] Bae S, Kim H, Lee Y, Xu X, Park J-S, Zheng Y, Balakrishnan J, Lei T, Ri Kim H, Song YI, Kim Y-J, Kim KS, Özyilmaz B, Ahn J-H, Hong BH, Iijima S. Roll-to-roll production of 30-inch graphene films for transparent electrodes. Nature Nanotechnology. 2010;**5**:574. DOI: 10.1038/nnano.2010.132

[22] Kim KS, Zhao Y, Jang H, Lee SY, Kim JM, Kim KS, Ahn J-H, Kim P, Choi J-Y, Hong BH. Large-scale pattern growth of graphene films for stretchable transparent electrodes. Nature. 2009;**457**:706. DOI: 10.1038/nature07719

[23] Wu J, Agrawal M, Becerril HA, Bao Z, Liu Z, Chen Y, Peumans P. Organic light-emitting diodes on solution-processed graphene transparent electrodes. ACS Nano. 2010;**4**:43-48. DOI: 10.1021/nn900728d

[24] Pang S, Hernandez Y, Feng X, Müllen K. Graphene as transparent electrode material for organic electronics. Advanced Materials. 2011;**23**:2779-2795. DOI: 10.1002/adma.201100304

[25] Lin Y-M, Dimitrakopoulos C, Jenkins KA, Farmer DB, Chiu H-Y, Grill A, Avouris P. 100-GHz transistors from wafer-scale epitaxial graphene. Science. 2010;**327**:662-662. DOI: 10.1126/science.1184289

[26] Chen F, Xia J, Ferry DK, Tao N. Dielectric screening enhanced performance in graphene FET. Nano Letters. 2009;**9**:2571-2574. DOI: 10.1021/nl900725u

[27] Fiori G, Iannaccone G. On the possibility of tunable-gap bilayer graphene FET. IEEE Electron Device Letters. 2009;**30**:261-264. DOI: 10.1109/led.2008.2010629

[28] Cho B, Yoon J, Hahm MG, Kim D-H, Kim AR, Kahng YH, Park S-W, Lee Y-J, Park S-G, Kwon J-D, Kim CS, Song M,

Jeong Y, Nam K-S, Ko HC. Graphene-based gas sensor: Metal decoration effect and application to a flexible device. Journal of Materials Chemistry C. 2014;**2**:5280-5285. DOI: 10.1039/c4tc00510d

[29] Sumboja A, Foo Ce Y, Wang X, Lee Pooi S. Large areal mass, flexible and free-standing reduced graphene oxide/manganese dioxide paper for asymmetric supercapacitor device. Advanced Materials. 2013;**25**:2809-2815. DOI: 10.1002/adma.201205064

[30] El-Kady MF, Strong V, Dubin S, Kaner RB. Laser scribing of high-performance and flexible graphene-based electrochemical capacitors. Science. 2012;**335**:1326-1330. DOI: 10.1126/science.1216744

[31] Feng L, Wang K, Zhang X, Sun X, Li C, Ge X, Ma Y. Flexible solid-state supercapacitors with enhanced performance from hierarchically graphene nanocomposite electrodes and ionic liquid incorporated gel polymer electrolyte. Advanced Functional Materials. 2017;**28**:1704463. DOI: 10.1002/adfm.201704463

[32] Glover CF, Richards CAJ, Williams G, McMurray HN. Evaluation of multi-layered graphene nano-platelet composite coatings for corrosion control part II—Cathodic delamination kinetics. Corrosion Science. 2018;**136**:304-310. DOI: 10.1016/j.corsci.2018.03.014

[33] Lee J, Berman D. Inhibitor or promoter: Insights on the corrosion evolution in a graphene protected surface. Carbon. 2018;**126**:225-231. DOI: 10.1016/j.carbon.2017.10.022

[34] Yu F, Camilli L, Wang T, Mackenzie DMA, Curioni M, Akid R, Bøggild P. Complete long-term corrosion protection with chemical vapor deposited graphene. Carbon. 2018;**132**:78-84. DOI: 10.1016/j.carbon.2018.02.035

[35] Qiu B, Xing M, Zhang J. Recent advances in three-dimensional graphene based materials for catalysis applications. Chemical Society Reviews. 2018;**47**:2165-2216. DOI: 10.1039/c7cs00904f

[36] Xi J, Wang Q, Liu J, Huan L, He Z, Qiu Y, Zhang J, Tang C, Xiao J, Wang S. N,P-dual-doped multilayer graphene as an efficient carbocatalyst for nitroarene reduction: A mechanistic study of metal-free catalysis. Journal of Catalysis. 2018;**359**:233-241. DOI: 10.1016/j.jcat.2018.01.003

[37] Varela AS, Kroschel M, Leonard ND, Ju W, Steinberg J, Bagger A, Rossmeisl J, Strasser P. pH effects on the selectivity of the electrocatalytic CO_2 reduction on graphene-embedded Fe–N–C motifs: Bridging concepts between molecular homogeneous and solid-state heterogeneous catalysis. ACS Energy Letters. 2018;**3**:812-817. DOI: 10.1021/acsenergylett.8b00273

[38] Wang B, Ryu J, Choi S, Song G, Hong D, Hwang C, Chen X, Wang B, Li W, Song H-K, Park S, Ruoff RS. Folding graphene film yields high areal energy storage in lithium-ion batteries. ACS Nano. 2018;**12**:1739-1746. DOI: 10.1021/acsnano.7b08489

[39] Geng P, Zheng S, Tang H, Zhu R, Zhang L, Cao S, Xue H, Pang H. Transition metal sulfides based on graphene for electrochemical energy storage. Advanced Energy Materials. 2018;**8**:1703259. DOI: 10.1002/aenm.201703259

[40] Mao J, Iocozzia J, Huang J, Meng K, Lai Y, Lin Z. Graphene aerogels for efficient energy storage and conversion. Energy & Environmental Science. 2018;**11**:772-799. DOI: 10.1039/c7ee03031b

[41] Lee J, Li Y, Tang JN, Cui XL. Synthesis of hydrogen substituted graphene through mechanochemistry

and its electrocatalytic properties. Acta Physico-Chimica Sinica. 2018;**34**:1080-1087. DOI: 10.3866/Pku. Whxb201802262

[42] Fu X, Hu Y, Zhang T, Chen S. The role of ball milled h-BN in the enhanced photocatalytic activity: A study based on the model of ZnO. Applied Surface Science. 2013;**280**:828-835. DOI: 10.1016/j.apsusc.2013.05.069

[43] Fu X, Hu Y, Yang Y, Liu W, Chen S. Ball milled h-BN: An efficient holes transfer promoter to enhance the photocatalytic performance of TiO_2. Journal of Hazardous materials. 2013;**244-245**:102-110. DOI: 10.1016/j.jhazmat.2012.11.033

[44] Fu X, Hu Y, Yang Y, Liu W, Chen S. Ball milled h-BN: An efficient holes transfer promoter to enhance the photocatalytic performance of TiO_2. Journal of Hazardous Materials. 2013;**244-245**:102-110. DOI: 10.1016/j.jhazmat.2012.11.033

[45] Shkodich NF, Vadchenko SG, Nepapushev AA, Kovalev DY, Kovalev ID, Ruvimov S, Rogachev AS, Mukasyan AS. Crystallization of amorphous $Cu_{50}Ti_{50}$ alloy prepared by high-energy ball milling. Journal of Alloys and Compounds. 2018;**741**:575-579. DOI: 10.1016/j.jallcom.2018.01.062

[46] Esquivel J, Wachowiak MG, O'Brien SP, Gupta RK. Thermal stability of nanocrystalline Al-5at.%Ni and Al-5at.%V alloys produced by high-energy ball milling. Journal of Alloys and Compounds. 2018;**744**:651-657. DOI: 10.1016/j.jallcom.2018.02.144

[47] Tie X, Han Q, Liang C, Li B, Zai J, Qian X. Si@SiO_x/graphene nanosheets composite: Ball milling synthesis and enhanced lithium storage performance. Frontiers in Materials. 2018;**4**:47. DOI: 10.3389/fmats.2017.00047

[48] Sun D, Ye D, Liu P, Tang Y, Guo J, Wang L, Wang H. MoS_2/graphene nanosheets from commercial bulky MoS_2 and graphite as anode materials for high rate sodium-ion batteries. Advanced Energy Materials. 2017;**8**:1702383. DOI: 10.1002/aenm.201702383

[49] Ma N, Jiang XY, Zhang L, Wang XS, Cao YL, Zhang XZ. Novel 2D layered molybdenum ditelluride encapsulated in few-layer graphene as high-performance anode for lithium-ion batteries. Small. 2018;**14**:1703680. DOI: 10.1002/smll.201703680

[50] Sha L, Gao P, Ren X, Chi Q, Chen Y, Yang P. A self-repairing cathode material for Lithium–selenium batteries: Se-C chemically bonded selenium-graphene composite. Chemistry—A European Journal. 2018;**24**:2151-2156. DOI: 10.1002/chem.201704079

[51] Chen J, Xu M-W, Wu J, Li CM. Center-iodized graphene as an advanced anode material to significantly boost the performance of lithium-ion batteries. Nanoscale. 2018;**10**:9115-9122. DOI: 10.1039/c8nr00061a

[52] Xia J, Liu L, Xie J, Yan H, Yuan Y, Chen M, Huang C, Zhang Y, Nie S, Wang X. Layer-by-layered SnS_2/graphene hybrid nanosheets via ball-milling as promising anode materials for lithium ion batteries. Electrochimica Acta. 2018;**269**:452-461. DOI: 10.1016/j.electacta.2018.03.022

[53] Zhu J, Zhou Y, Wang B, Zheng J, Ji S, Yao H, Luo H, Jin P. Vanadium dioxide nanoparticle-based thermochromic smart coating: High luminous transmittance, excellent solar regulation efficiency, and near room temperature phase transition. ACS Applied Materials & interfaces. 2015;**7**:27796-27803. DOI: 10.1021/acsami.5b09011

[54] Zhong L, Li M, Wang H, Luo Y, Pan J, Li G. Star-shaped VO_2 (M) nanoparticle films with high thermochromic performance. CrystEngComm. 2015;**17**:5614-5619. DOI: 10.1039/c5ce00873e

[55] Zhang J, Jin H, Chen Z, Cao M, Chen P, Dou Y, Zhao Y, Li J. Self-assembling VO_2 nanonet with high switching performance at wafer-scale. Chemistry of Materials. 2015;**27**:7419-7424. DOI: 10.1021/acs.chemmater.5b03314

[56] Zhang H, Li Q, Shen P, Dong Q, Liu B, Liu R, Cui T, Liu B. The structural phase transition process of free-standing monoclinic vanadium dioxide micron-sized rods: Temperature-dependent Raman study. RSC Advances. 2015;**5**:83139-83143. DOI: 10.1039/c5ra15947d

[57] Powell MJ, Marchand P, Denis CJ, Bear JC, Darr JA, Parkin IP. Direct and continuous synthesis of VO_2 nanoparticles. Nanoscale. 2015;**7**:18686-18693. DOI: 10.1039/c5nr04444h

[58] Wang R, Han M, Zhao Q, Ren Z, Guo X, Xu C, Hu N, Lu L. Hydrothermal synthesis of nanostructured graphene/polyaniline composites as high-capacitance electrode materials for supercapacitors. Scientific Reports. 2017;**7**:44562

[59] Nathan DMGT, Boby SJM. Hydrothermal preparation of hematite nanotubes/reduced graphene oxide nanocomposites as electrode material for high performance supercapacitors. Journal of Alloys and Compounds. 2017;**700**:67-74. DOI: 10.1016/j.jallcom.2017.01.070

[60] Meng S, Ye X, Ning X, Xie M, Fu X, Chen S. Selective oxidation of aromatic alcohols to aromatic aldehydes by BN/metal sulfide with enhanced photocatalytic activity. Applied Catalysis B: Environmental. 2016;**182**:356-368. DOI: 10.1016/j.apcatb.2015.09.030

[61] Alie D, Gedvilas L, Wang Z, Tenent R, Engtrakul C, Yan Y, Shaheen SE, Dillon AC, Ban C. Direct synthesis of thermochromic VO_2 through hydrothermal reaction. Journal of Solid State Chemistry. 2014;**212**:237-241. DOI: 10.1016/j.jssc.2013.10.023

[62] Pang Q, Zhao Y, Yu Y, Bian X, Wang X, Wei Y, Gao Y, Chen G. VS_4 nanoparticles anchored on graphene sheets as a high-rate and stable electrode material for sodium ion batteries. ChemSusChem. 2018;**11**:735-742. DOI: 10.1002/cssc.201702031

[63] Zou Y, Zhang Z, Zhong W, Yang W. Hydrothermal direct synthesis of polyaniline, graphene/polyaniline and N-doped graphene/polyaniline hydrogels for high performance flexible supercapacitors. Journal of Materials Chemistry A. 2018;**6**:9245-9256. DOI: 10.1039/c8ta01366g

[64] Lee JS, You KH, Park CB. Highly photoactive, low bandgap TiO_2 nanoparticles wrapped by graphene. Advanced Materials. 2012;**24**: 1084-1088. DOI: 10.1002/adma.201104110

[65] Liu C, Sun H, Qian J, Chen Z, Chen F, Liu S, Lv Y, Lu X, Chen A. Ultrafine Mn_3O_4/CeO_2 nanorods grown on reduced graphene oxide sheets as high-performance supercapacitor electrodes. Journal of Alloys and Compounds. 2017;**722**:54-59. DOI: 10.1016/j.jallcom.2017.06.097

[66] Lee JW, Hall AS, Kim J-D, Mallouk TE. A facile and template-free hydrothermal synthesis of Mn_3O_4 nanorods on graphene sheets for supercapacitor electrodes with long cycle stability. Chemistry of Materials. 2012;**24**:1158-1164. DOI: 10.1021/cm203697w

[67] Mantilaka MMMGPG, De Silva RT, Ratnayake SP, Amaratunga G, de Silva KMN. Photocatalytic activity of electrospun MgO nanofibres: Synthesis, characterization and applications. Materials Research Bulletin. 2018;**99**:204-210. DOI: 10.1016/j.materresbull.2017.10.047

[68] Zhang L, Zhang Q, Xie H, Guo J, Lyu H, Li Y, Sun Z, Wang H, Guo Z. Electrospun titania nanofibers segregated by graphene oxide for improved visible light photocatalysis. Applied Catalysis B: Environmental. 2017;**201**:470-478. DOI: 10.1016/j.apcatb.2016.08.056

[69] Han C, Mao W, Bao K, Xie H, Jia Z, Ye L. Preparation of Ag/Ga$_2$O$_3$ nanofibers via electrospinning and enhanced photocatalytic hydrogen evolution. International Journal of Hydrogen Energy. 2017;**42**:19913-19919. DOI: 10.1016/j.ijhydene.2017.06.076

[70] Zhang Y, Park M, Kim HY, Ding B, Park S-J. In-situ synthesis of nanofibers with various ratios of BiOCl$_x$/BiOBr$_y$/BiOI$_z$ for effective trichloroethylene photocatalytic degradation. Applied Surface Science. 2016;**384**:192-199. DOI: 10.1016/j.apsusc.2016.05.039

[71] Zhang Y, Park M, Kim H-Y, El-Newehy M, Rhee KY, Park S-J. Effect of TiO$_2$ on photocatalytic activity of polyvinylpyrrolidone fabricated via electrospinning. Composites Part B: Engineering. 2015;**80**:355-360. DOI: 10.1016/j.compositesb.2015.05.040

[72] Wang K, Shao C, Li X, Zhang X, Lu N, Miao F, Liu Y. Hierarchical heterostructures of p-type BiOCl nanosheets on electrospun n-type TiO$_2$ nanofibers with enhanced photocatalytic activity. Catalysis Communications. 2015;**67**:6-10. DOI: 10.1016/j.catcom.2015.03.037

[73] Kayaci F, Vempati S, Ozgit-Akgun C, Donmez I, Biyikli N, Uyar T. Selective isolation of the electron or hole in photocatalysis: ZnO-TiO$_2$ and TiO$_2$-ZnO core-shell structured heterojunction nanofibers via electrospinning and atomic layer deposition. Nanoscale. 2014;**6**:5735-5745. DOI: 10.1039/c3nr06665g

[74] Wang H, Wang W, Wang H, Jin X, Niu H, Wang H, Zhou H, Lin T. High performance supercapacitor electrode materials from electrospun carbon nanofibers in situ activated by high decomposition temperature polymer. ACS Applied Energy Materials. 2018;**1**:431-439. DOI: 10.1021/acsaem.7b00083

[75] Ekabutr P, Klinkajon W, Sangsanoh P, Chailapakul O, Niamlang P, Khampieng T, Supaphol P. Electrospinning: A carbonized gold/graphene/PAN nanofiber for high performance biosensing. Analytical Methods. 2018;**10**:874-883. DOI: 10.1039/c7ay02880f

[76] Mishra RK, Nawaz MH, Hayat A, Nawaz MAH, Sharma V, Marty J-L. Electrospinning of graphene-oxide onto screen printed electrodes for heavy metal biosensor. Sensors and Actuators B: Chemical. 2017;**247**:366-373. DOI: 10.1016/j.snb.2017.03.059

[77] Chee WK, Lim HN, Zainal Z, Harrison I, Andou Y, Huang NM, Altarawneh M, Jiang ZT. Electrospun graphene nanoplatelets-reinforced carbon nanofibers as potential supercapacitor electrode. Materials Letters. 2017;**199**:200-203. DOI: 10.1016/j.matlet.2017.04.086

[78] Wei M, Jiang M, Liu X, Wang M, Mu S. Graphene-doped electrospun nanofiber membrane electrodes and proton exchange membrane fuel cell performance. Journal of Power Sources. 2016;**327**:384-393. DOI: 10.1016/j.jpowsour.2016.07.083

[79] Schütz MB, Xiao L, Lehnen T, Fischer T, Mathur S. Microwave-assisted synthesis of nanocrystalline binary and ternary metal oxides. International Materials Reviews. 2018;**63**:341-374. DOI: 10.1080/09506608.2017.1402158

[80] Zhong Y, Yu L, Chen Z-F, He H, Ye F, Cheng G, Zhang Q. Microwave-assisted synthesis of Fe$_3$O$_4$ nanocrystals with predominantly exposed facets and their heterogeneous UVA/Fenton

catalytic activity. ACS Applied Materials & Interfaces. 2017;**9**:29203-29212. DOI: 10.1021/acsami.7b06925

[81] Kumar R, da Silva ET, Singh RK, Savu R, Alaferdov AV, Fonseca LC, Carossi LC, Singh A, Khandka S, Kar KK. Microwave-assisted synthesis of palladium nanoparticles intercalated nitrogen doped reduced graphene oxide and their electrocatalytic activity for direct-ethanol fuel cells. Journal of Colloid and Interface Science. 2018;**515**:160-171. DOI: 10.1016/j.jcis.2018.01.028

[82] Shi S, Deng T, Zhang M, Yang G. Fast facile synthesis of SnO_2/graphene composite assisted by microwave as anode material for lithium-ion batteries. Electrochimica Acta. 2017;**246**:1104-1111. DOI: 10.1016/j.electacta.2017.06.111

[83] Dursun B, Topac E, Alibeyli R, Ata A, Ozturk O, Demir-Cakan R. Fast microwave synthesis of SnO_2@graphene/N-doped carbons as anode materials in sodium ion batteries. Journal of Alloys and Compounds. 2017;**728**:1305-1314. DOI: 10.1016/j.jallcom.2017.09.081

[84] Wang N, Goh QS, Lee PL, Magdassi S, Long Y. One-step hydrothermal synthesis of rare earth/W-codoped VO_2 nanoparticles: Reduced phase transition temperature and improved thermochromic properties. Journal of Alloys and Compounds. 2017;**711**:222-228. DOI: 10.1016/j.jallcom.2017.04.012

[85] Wang N, Duchamp M, Dunin-Borkowski RE, Liu S, Zeng X, Cao X, Long Y. Terbium-doped VO_2 thin films: Reduced phase transition temperature and largely enhanced luminous transmittance. Langmuir. 2016;**32**:759-764. DOI: 10.1021/acs.langmuir.5b04212

[86] Wang N, Duchamp M, Xue C, Dunin-Borkowski RE, Liu G, Long Y. Single-crystalline W-doped VO_2 nanobeams with highly reversible electrical and plasmonic responses near room temperature. Advanced Materials Interfaces. 2016;**3**:1600164. DOI: 10.1002/admi.201600164

[87] Wang N, Liu S, Zeng XT, Magdassi S, Long Y. Mg/W-codoped vanadium dioxide thin films with enhanced visible transmittance and low phase transition temperature. Journal of Materials Chemistry C. 2015;**3**:6771-6777. DOI: 10.1039/c5tc01062d

[88] Mounasamy V, Mani GK, Ponnusamy D, Tsuchiya K, Prasad AK, Madanagurusamy S. Template-free synthesis of vanadium sesquioxide (V_2O_3) nanosheets and their room-temperature sensing performance. Journal of Materials Chemistry A. 2018;**6**:6402-6413. DOI: 10.1039/c7ta10159g

[89] Alexeev AM, Barnes MD, Nagareddy VK, Craciun MF, Wright CD. A simple process for the fabrication of large-area CVD graphene based devices via selective in situ functionalization and patterning. 2d Materials. 2017;**4**:011010. DOI: 10.1088/2053-1583/4/1/011010

[90] Yang Y, Lee K, Zobel M, Mackovic M, Unruh T, Spiecker E, Schmuki P. Formation of highly ordered VO_2 nanotubular/nanoporous layers and their supercooling effect in phase transitions. Advanced Materials. 2012;**24**:1571-1575. DOI: 10.1002/adma.201200073

Chapter 5

Graphene Acoustic Devices

He Tian, Guang-Yang Gou, Fan Wu, Lu-Qi Tao, Yi Yang and Tian-Ling Ren

Abstract

In 2011, Ren's group has developed the first graphene sound source device in the world. This is the first time that the graphene applications have been extended into acoustic area. The graphene sound source can produce sound in a wide sound frequency range from 100 Hz to 50 kHz. After that, we have innovated the first graphene earphone, which can be used both for human and animals. In 2017, both the sound detection and sound emission have been integrated into one graphene device, which is called graphene artificial throat. In this book chapter, more details for developing those graphene acoustic devices will be introduced, which can help to boost the real applications of graphene devices.

Keywords: graphene sound, acoustic, thermoacoustic effect, wide frequency range, flexible, large-scale

1. Introduction

Since graphene have been found in 2004 [1], various kinds of graphene-based devices, including field effect transistor (FET) [2], memory [3], photodetector [4], sensor [5], have been built as its excellent structural and physical properties. However, almost no work was focused on the acoustic device in audio range because the graphene is hard to make low frequency sound through vibration due to the large-area requirement. Although the working principles of traditional sound devices are different, they are all depend on mechanical vibration of thin films, and driving the air to produce the sound. There is a common problem that the output audio spectrum of these sound devices is not flat, which is caused by the inherent center resonance of the diaphragm. Here, thermoacoustic effect is proposed to emit sound without vibration of the diagram. The conductive film itself can emit sound, which will be expected to achieve wide-band acoustic output. In 1917, Arnold's group firstly used the 700 nm-Pt film as the source to realize thermoacoustic sound production [6]. The sound frequency of this device can reach to 40 kHZ, but the sound pressure (SP) was not high enough. With the rapid development of nanotechnology in recent years, many outstanding progresses have been made in sound source devices based on thermoacoustic effect. In 1999, Shinoda's group have reported aluminum thin film sound device based on porous silicon substrate, and the wide output band can be implemented in 20–100 kHz [7]. Especially, the SP of the device is up to 0.1 Pa. After that, the research group in Tsinghua University realized the sound source device based on carbon nanotubes [8]. This thermoacoustic device not only can produce high sound pressure level (SPL) in a wide audio range, but also have the advantages of bending, stretching and transparency.

However, most of thermoacoustic devices have some problems and technical limitations in material preparation, device structure and working mechanism. For example, the performance of aluminum thin film thermoacoustic device is seriously degraded because the aluminum is easily oxidized in air, and the device is rigid and non-transparent [7]. For the thermoacoustic device based on carbon nanotubes [8], 100 V is needed to drive the device due to its large square resistance (1 kΩ/$\square$).

For the currently existing problems of low reliability, poor performance and high driving voltage, the high-performance thermoacoustic device must meet three conditions. First, the conductor should be thin enough with a low thermal capacity per unit area (HCPUA). Second, the conductor should prevent thermal leakage from the substrate (i.e. suspend). Third, the conductor area should be large enough to build a sufficient sound field. Graphene as a kind of two-dimensional layered material provides an opportunity for the development of thermoacoustic devices due to its ultralow HCPUA. After a lot of efforts, Ren's group has made some new achievements in graphene thermoacoustic devices. In particular, their graphene earphones and graphene throat have attracted widespread attention as their potential applications in solving the problem of human listening and speaking. Therefore, this chapter will detail the graphene sound sources, and explore the method of realization high performance devices.

2. Establishment of theoretical model

The physical image of thermoacoustic effect can be described as: when an alternating current signal is passing through the thin film, a film Joule heat is generated and quickly transferred to the surrounding air medium. Due to the periodic rise and fall of the surface temperature, a thin layer of air molecules on the film surface will continuously expand and contract to produce sound waves. **Figure 1a** shows the detailed physical process of the thermoacoustic device. The input AC signal (electric energy) is converted into Joule thermal fluctuation (thermal energy), and finally converted into sound waves (sound energy). Different from the principle of traditional acoustic devices, the conductive film itself does not vibrate itself but

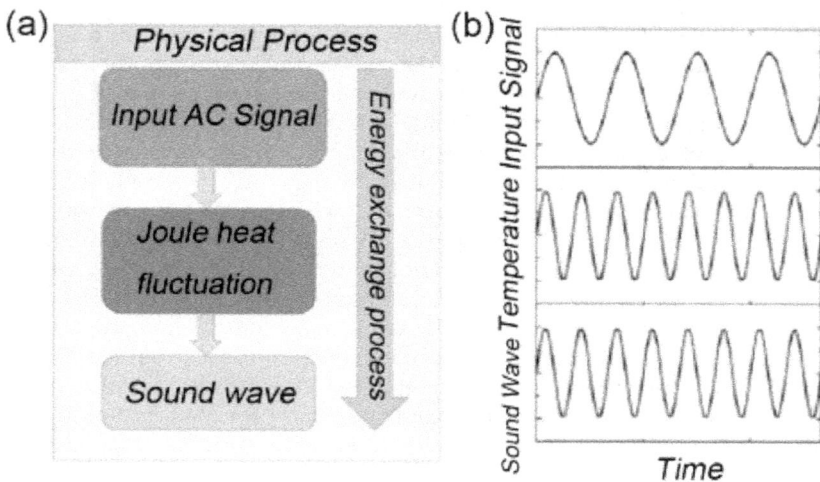

Figure 1.
(a) Energy conversion process of thermoacoustic devices. (b) Theoretical waveforms during thermoacoustic effect [9].

makes the air vibrate by heating the air medium during the process of thermoacoustic effect. **Figure 1b** shows the theoretical waveforms corresponding to the three kinds of energy in the time domain. Assuming a sinusoidal signal is input, the Joule heat generated is squared with the electrical signal. Both positive half-cycle and negative half-cycle electrical signals can produce positive temperature fluctuations. Therefore, the frequency of temperature changes is twice the frequency of the input electrical signal, and the frequency of the sound waves is also doubled. This is an important feature that distinguishes the thermoacoustic effect from the traditional sound production.

In order to obtain high performance graphene sound source devices, it is necessary to carry out theoretical design to guide the experiment. Firstly, the theoretical model of thermoacoustic effect is established, and the sound pressure produced by graphene is predicted theoretically. The structure of graphene/substrate/back plate is proposed in **Figure 2**. The heat loss from the substrate should be taken into account due to the contact of graphene with the substrate. When the acoustic frequency is low, the heat flux can penetrate the substrate to reach the backplane, but when the sound frequency is high, the heat flux cannot reach the backplane. So it needs to be discussed in two frequency bands [9–11]:

when $f < \frac{\alpha_s}{4\pi L_S^2}$, the SP generated by graphene in the far field can be expressed as:

$$P_{rms} = \frac{\gamma - 1}{\sqrt{2}\,v_g} \frac{e_g}{M(e_s + a_c) + e_g} q_0 \tag{1}$$

when $f > \frac{\alpha_s}{4\pi L_S^2}$, the SP generated by graphene in the far field can be expressed as:

$$P_{rms} = \frac{\gamma - 1}{\sqrt{2}\,v_g} \frac{e_g}{(e_s + a_c) + e_g} q_0 \tag{2}$$

where f is frequency of sound; α_s and L_s are the thermal diffusivity and thickness of substrate, respectively; γ is the heat capacity ratio of gas; v_g is the sound velocity in gas; each layer has $e_i = \sqrt{K_i \rho_i C_{p,i}}$ is the thermal effusivity of material i. The

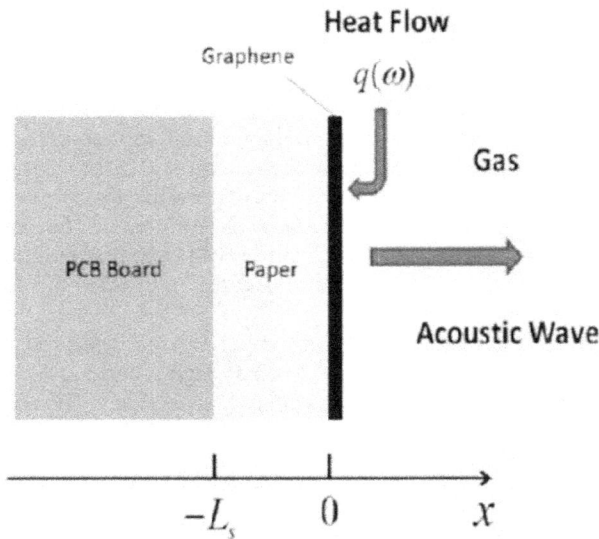

Figure 2.
Theoretical model of the graphene sound source [10].

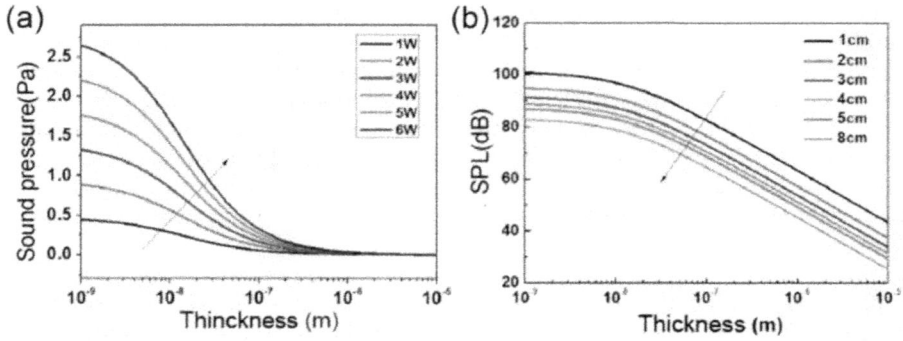

Figure 3.
(a) Theoretical relationship between SP and thickness of graphene under different input power conditions.
(b) Theoretical relationship between graphene SP and thickness under different test distance conditions [8].

subscript i represent gas (g), substrate (s), or backing layer (b), respectively. q_0 is the input power density; k_i, ρ_i and $C_{p,i}$ are the thermal conductivity, density and specific heat capacity of each layer of material, respectively; M is a frequency related factor. Under high frequency, $M \approx 1$, then the two equations are the same.

Based on the above theoretical formula, the relationship between the SP and thickness of graphene, input power and test distance can be calculated and analyzed. **Figure 3a** shows the theoretical waveform of SP. It can been seen that the corresponding SP value increases with the thickness of graphene decreasing, and the SP will increase as the input power increases at the same thickness. **Figure 3b** shows the theoretical relationship between SPL and graphene thickness at different test distances. At the same thickness, the SPL value will decrease as the test distance increases.

Then the sound field radiation of graphene sound source device is analyzed. The sound source device can be regarded as a point sound source in the far field, and the theoretical acoustic directivity $D(\theta, \varphi)$ can be expressed as [12]:

$$D(\theta, \varphi) = \text{sinc}\left(\frac{k_0 L_x}{2} \sin\theta \cos\varphi\right) \text{sinc}\left(\frac{k_0 L_y}{2} \sin\theta \sin\varphi\right) \qquad (3)$$

where θ and φ are the parameters of the spherical coordinate system; $k_0 = 2\pi/\lambda_0$ is the wave number; Assuming that the sound source device is at the center of the sound field, L_x and L_y are the length and width of the sound source, respectively. **Figure 4** shows the directivity of the graphene point sound source at different sound frequencies. When the sound frequency is lower, the corresponding acoustic wave is longer, and the angle of sound field coverage is lower. With the increase of acoustic frequency, the acoustic wavelength decreases, and the sound field becomes more concentrated in the smaller angle perpendicular to the sample direction, that is, the directivity is enhanced.

Then, the sound field distribution of graphene device is analyzed, which includes near field and far field. Firstly, the distance of Rayleigh is defined as A/λ, where A and λ are the area and wavelength of graphene, respectively. The acoustic wave propagates in the form of plane wave when the test distance is smaller than Rayleigh distance in near field. However, when the measured distance is larger than Rayleigh distance, the acoustic wave propagates in the form of spherical waves in far field. For $r < R_0$, the SP in near field can be expressed as:

$$P(\theta, \varphi) = \text{sinc}\left(\frac{k_0 L_x}{2} \sin\theta \cos\varphi\right) \text{sinc}\left(\frac{k_0 L_y}{2} \sin\theta \sin\varphi\right) \qquad (4)$$

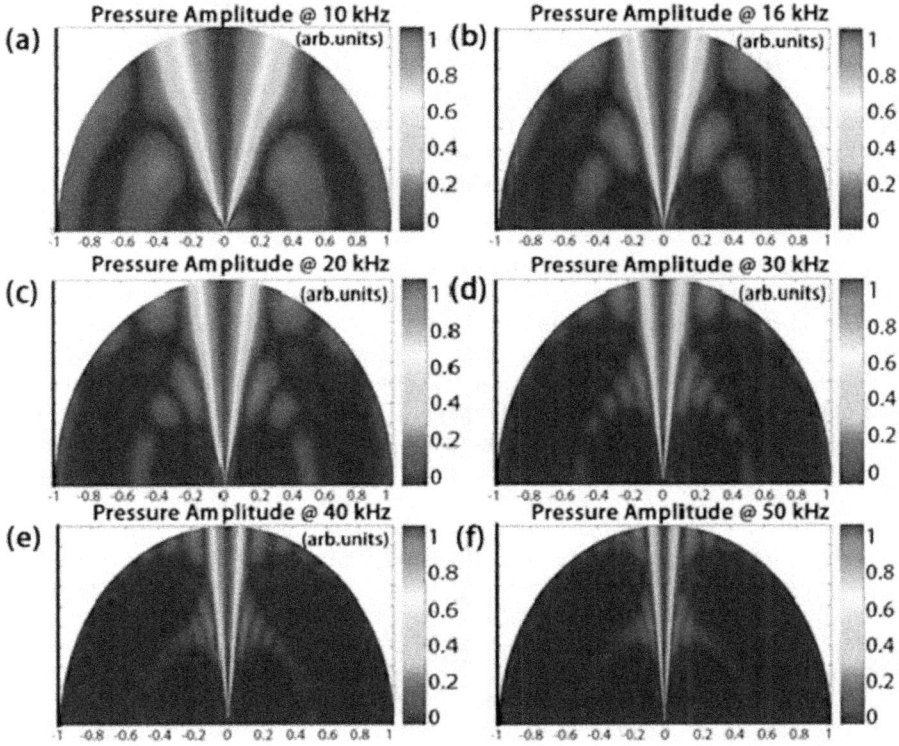

Figure 4.
Directivity distribution of graphene sound source at different sound frequencies [9]. (a) 10 kHz, (b) 16 kHz, (c) 20 kHz, (d) 30 kHz, (e) 40 kHz, (f) 50 kHz.

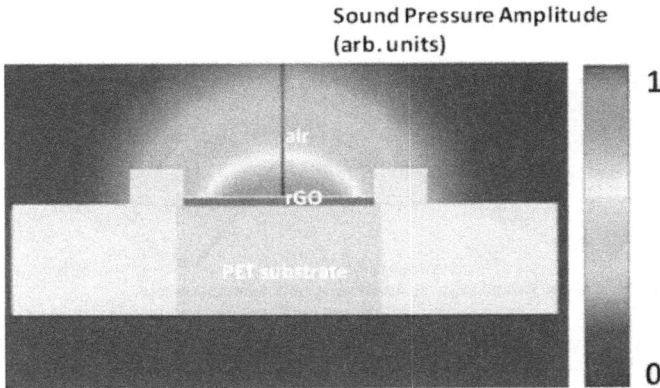

Figure 5.
SP distribution of graphene sound source device obtained by finite element simulation [9].

For $r > R_0$, the SP in far field can be expressed as:

$$P(\theta, \varphi) = \frac{R_0}{r} \mathrm{sinc}\left(\frac{k_0 L_x}{2} \sin\theta \cos\varphi\right) \mathrm{sinc}\left(\frac{k_0 L_y}{2} \sin\theta \sin\varphi\right) \qquad (5)$$

Based on the above two formulas, the theoretical distribution of graphene sound field can be simulated.

In order to fully consider the influence of the actual size of the device on the sound field distribution, the finite element software Comsol is used to simulate the sound field distribution. **Figure 5** shows the sound field distribution of graphene sound source devices simulated by finite element software. The simulation results show that the SP distribution on the surface of the device is stronger and the angle of sound coverage is wider. However, the sound coverage angle obtained by simulation is narrower when the point sound source approximation is used (**Figure 4**).

3. Graphene-on-paper sound source devices

The fabrication process of the multilayer graphene sound source device is introduced in **Figure 6**. Firstly, multilayer graphene was prepared on nickel foil by CVD method, the thickness of which was about 20 nm. Then, the graphene film of 1 cm × 1 cm was transferred to filter paper. The transfer process was as follows: the $FeCl_3$ solution was used to etch graphene with nickel substrate. After nickel was etched, graphene was transferred to deionized water, and finally graphene was filled with porous filter paper. In the transfer process, filter paper with the pore size of 30–50 μm was used as substrate. Because the graphene film can be suspended on larger pores, which can effectively reduce heat loss to substrate and improve the efficiency of sound production. In order to test the sample acoustically, the sample was installed on the PCB board and the two electrodes were made. The Ag was used as electrode material of graphene sound source device. **Figure 6** shows the schematic diagram of a graphene-on-paper sound source device. When the sound frequency electric signal is applied to graphene device, the air near its surface can be heated, and then the periodicity of air vibration can induce sound waves.

Figure 7a shows the scanning electron microscope (SEM) image of the graphene. There are some ripples in graphene to get regular graphics. The oxygen plasma can be used to pattern graphene film. **Figure 7b** shows the image of the graphene film

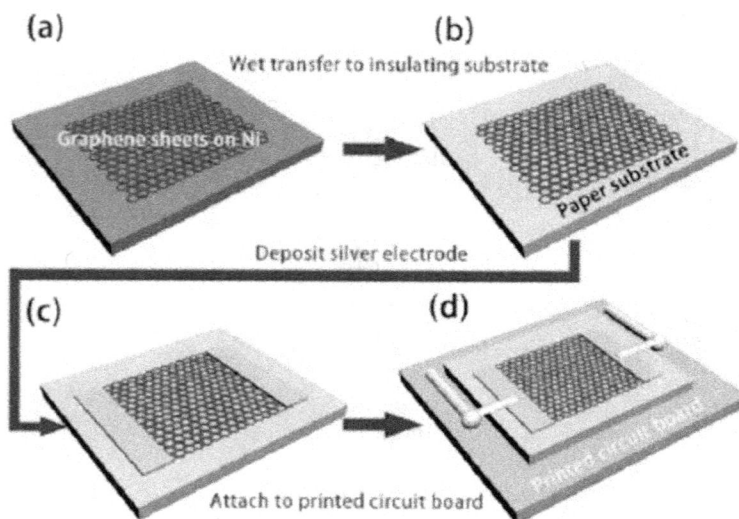

(a)　　　　　　　　　**(b)**

Wet transfer to insulating substrate

Graphene sheets on Ni

Paper substrate

Deposit silver electrode

(c)　　　　　　　　　**(d)**

Attach to printed circuit board

Figure 6.
A process for preparing a multilayer graphene sound source device: (a) growing a multilayer graphene on nickel; (b) transfer to filter paper by wet method; (c) prepare the Ag electrode at both ends of the graphene; (d) mount the device to the backplate and extract the signal [9].

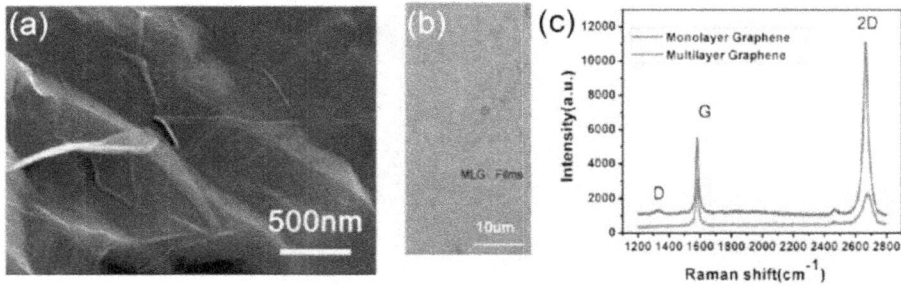

Figure 7.
(a) SEM image of graphene. (b) An optical image of the graphene after oxygen plasma etching. (c) Raman spectrum in the range of 1200–2800 cm^{-1} of the graphene. The red line and green line correspond to the red and green part line in **Figure 7b,** *respectively [9].*

after oxygen plasma treatment. **Figure 7c** shows the Raman peak of the sample obtained from the corresponding colored spots in **Figure 7b**.

It can be seen that the two strong peaks locate at 1582 and 2700 cm^{-1}, corresponding to G and 2D bands, respectively. The sample exhibits typical multilayered graphene characteristic with a strong G peak and a broad 2D peak (green line), while some spots exhibit monolayer graphene feature with a sharp G peak and a single higher 2D peak (red line). The small intensity of D-band observed at 1350 cm^{-1} indicates the low levels of defects and local-disorders in the deposited films.

Three paper-based graphene sound source devices are fabricated, named sample 1, sample 2 and sample 3, and their resistance are 32, 143, and 601 Ω, respectively. The area of these devices is 1 cm × 1 cm, and the average thickness of three graphene sheet samples are about 100, 60, and 20 nm, respectively. The acoustics test platform is composed of a signal generator, a standard microphone and a dynamic frequency analyzer, as shown in **Figure 8a**. The signal generator drives the graphene sound source device to produce sound, and the sound wave is received by the standard microphone. Finally, the sound wave is analyzed by the dynamic frequency analyzer and the frequency is converted from time domain to frequency. The graphene sound source is directly tested by using a standard microphone (**Figure 8b**). The distance from the sound source to the microphone is 5 cm. The relationship between output SP and the input power is shown in **Figure 8c**. The result indicates that the SP increases linearly with increasing input power. **Figure 8d** shows the relationship between the SP and the test distance. For the graphene sound source devices, the Rayleigh distance can be calculated as 4.7 × 10^{-3} m. The test distance of SP is ranging from 1 to 10 × 10^{-2} m, which belongs to the far-field. The SP decreases with the increase of the test distance at 16 kHz sound frequency, indicating an inverse proportional to distance. This result is in agreement with the estimate of far-field. When the measure distance is 5 cm at 16 kHz sound frequency, the omnidirectional dispersion patterns can be achieved by testing the change of the SP with the receiving angle. The sound field directivity of multilayer graphene is shown in **Figure 8e**. The SP is mainly concentrated within the ±30° of the positive axis. After this angle, the SP is obviously attenuated. **Figure 8f** shows the relation between the output SPL and the frequency. The frequency is ranging from 3 to 50 kHz. The three curves are normalized with the same power density (1 W/cm^2) at different thickness of the graphene film. It can be seen that the 20 nm graphene film has the highest SPL value because of its lowest HCPUA. The 60 and 100 nm graphene rank second and third, respectively. It indicates that thinner graphene sheets can produce higher SPL. The sound frequency band can cover audible and ultrasound. Especially in ultrasound range 20–50 kHz, there exists flat frequency response.

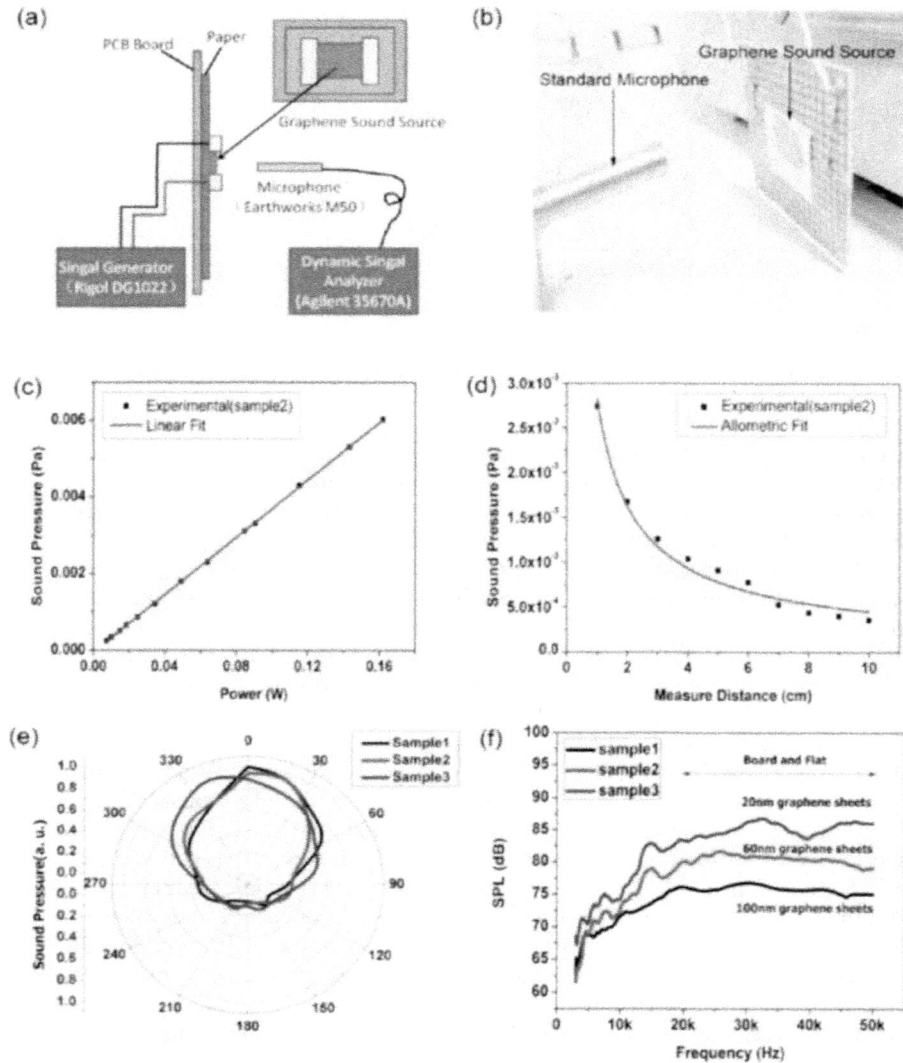

Figure 8.
The acoustic test platform and test results of graphene sound source [10]. (a) Schematic diagram of test platform. (b) Onsite photo of the experimental setup. (c) The output SP from graphene versus the input power. (d) The plot of the output SP of graphene versus the measurement distance. (e) Directivity of the graphene sound source in far-field. (f) The output SPL versus the frequency. The three curves are normalized with the input power 1 W/cm².

The theoretical model has been introduced in the previous section. **Figure 9** shows the sound radiation of the graphene sound source in far-field. The on-axis direction has the largest sound intensity, the sound intensity decreases with the angle and the main intensity area focuses on axis ±30 angles. Those results are agreed with the experimental directivity as shown in **Figure 8e**.

The physical scene of thermoacoustic effects is depicted in Arnold and Crandall's research results. When thermoacoustic device is working, alternating electrical signals generate joule heat through conductors. It would heat up the air near its surface and then the SP is generated by the changing of the air temperature. To verify this physical phenomenon through experiments, the advanced

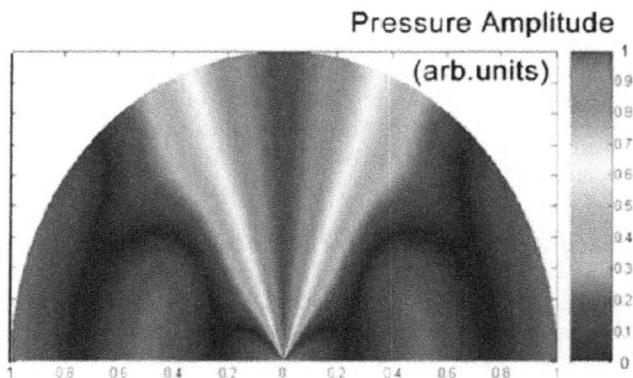

Figure 9.
Theoretical half-space directivity of the graphene sound source in far-field [10].

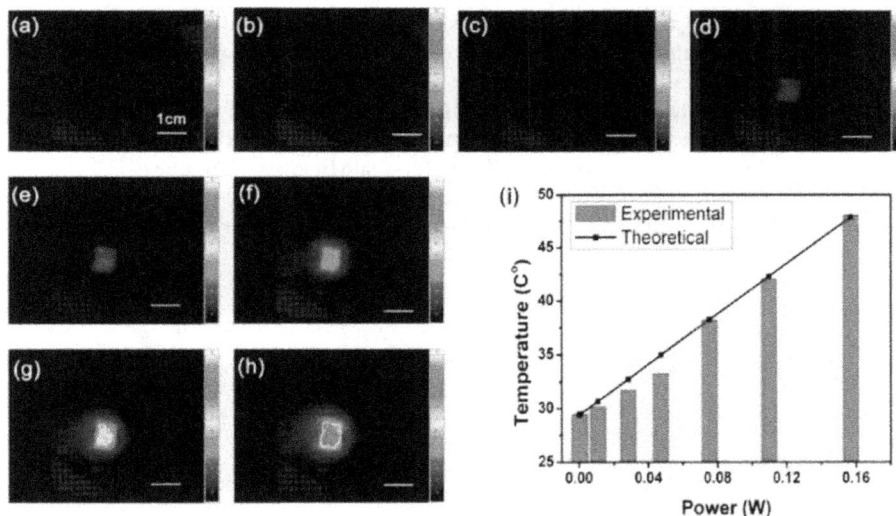

Figure 10.
Infrared thermal images and average surface temperature of graphene (sample 2) with different amplitude of input power [10] (a) no power is applied; (b) input power q_0 is 0.0007 W; (c) q_0 is 0.01 W; (d) q_0 is 0.03 W; (e) q_0 is 0.05 W; (f) q_0 is 0.08 W; (g) q_0 is 0.11 W; (h) q_0 is 0.16 W. (i) The average surface temperature of graphene versus the applied electric power. The experimental and theoretical results are shown. The SP is recorded at 16 kHz sound frequency and 5 cm measurement distance.

infrared thermal imaging instrument is used to investigate to the relations between the surface temperature distribution of graphene and the amplitude of input power. The input power q_0 is increasing from 0 to 0.16 W, the temperature distributions are collected in **Figure 10a–h**. The relationship between input power and graphene surface temperature is shown in **Figure 10i**. When the input power is 0 W, the surface temperature of the graphene is the same as that of room temperature. With the increase of input power, the surface temperature is gradually improved. The average surface temperature T of graphene can be expressed as [7]:

$$T = T_0 + \frac{q_0}{\sqrt{j\omega\, k_s\, C_{p,s}}} \qquad (6)$$

where k_s and $C_{p,s}$ are the thermal conductivity and heat capacity of the paper substrate, respectively; ω is the angular frequency of sound. The average temperature is linearly related to the input power, and the theoretical curve is in good agreement with the experimental results. Combined with the test results of **Figure 8c**, the SP increases with the input power, which indicates that the Joule heating is related to the SP and the working mechanism is thermoacoustic effect.

4. Graphene earphones: entertainment for both humans and animals

Laser scribing can be used to prepare graphene due to its advantage of low cost, high speed and no transfer. The earphone based on laser scribed graphene (LSG) can cover audible range and ultrasonic range. Animal hearing is more sensitive to the sound of the ultrasound band than audible domain. Therefore, the graphene earphones can be applied not only to humans, but also to animals. This section will introduce laser direct writing to prepare graphene earphones. The advantage of fabricating graphene earphones by using the laser scribing technology is that the large scale array can be prepared without the mask. **Figure 11a** shows the preparation process of graphene earphones. The graphene oxide (GO) solution can be directly coated on PET substrates. Then the GO film can be reduced to graphene under 788 nm laser irradiation by using the DVD light engraving machine. It is noted that the preparation of the wafer-scale precise graphene earphones requires only 25 minutes by using laser scribing technology. **Figure 11b** shows the wafer-scale flexible graphene earphones on the PET substrate. The inset in **Figure 11b** shows the graphene earphones at 1 cm^2 dimensions. The SEM image of graphene sheets is shown in **Figure 11c**. Before the

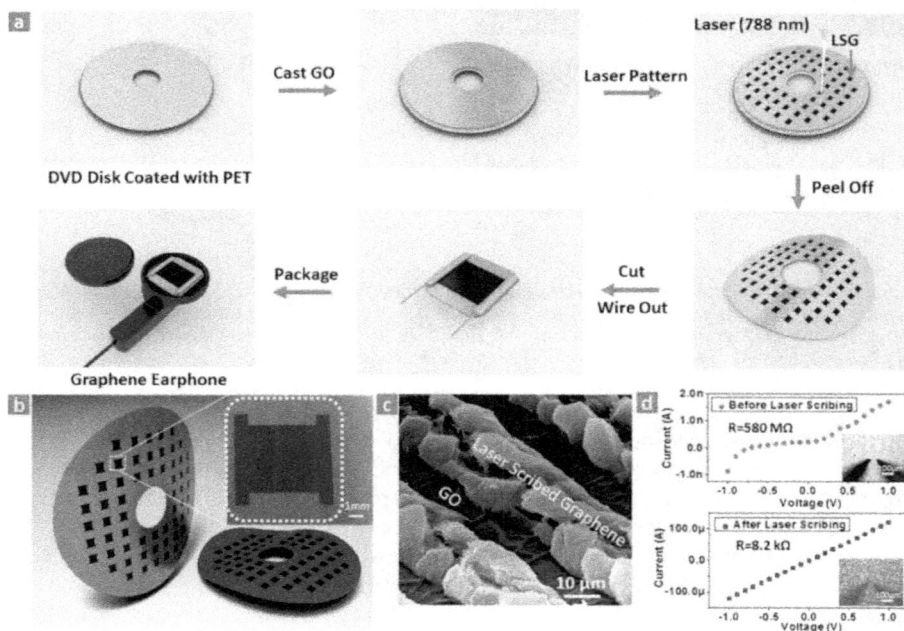

Figure 11.
Laser scribing technology for flexible graphene earphone fabrication [13]. (a) Process flow of the graphene earphone fabrication; (b) wafer-scale flexible graphene earphones. The inset shows an optical microscope image of the LSG earphone; (c) SEM image of the LSG and GO in false color; (d) electrical properties before and after laser scribing.

laser scribing, the GO film is quite flat and dense. After the laser scribing, a graphene film of 10 μm thickness is obtained. It is noticed that there is an almost 10-fold thickness increase for the LSG compared to the original GO film. **Figure 11d** shows the I-V curves of the film before and after the laser scribing. The resistance of the GO film is 580 MΩ. However, when the GO film is reduced to the LSG, the small resistance can be obtained as 8.2 kΩ, which is an almost five orders of magnitude reduction.

Graphene sound source devices can be packaged into traditional earphones. After the laser scribing of the wafer-scale graphene patterns, these patterns could be cut into individual graphene earphones. **Figure 12a** shows the graphene earphone with an area of 10 × 10 mm². Silver paste is applied on both sides of the graphene film to establish electrical input. The graphene earphones are finally connected to the electrical wiring of a commercial earphone casing using copper wires, as shown in **Figure 12b**. The structure of the graphene earphone is made up of the top cap, the graphene sheets, the Ag electrodes, the PET, and the bottom cap, as shown in **Figure 12c**. The packaged graphene earphones for human use is shown in **Figure 12d**. Compared with the traditional earphones, the distinctive feature of graphene device is significantly thinner than a voice coil. In order to obtain a sufficiently high SPL and equal-frequency playback sound, the periphery circuit is successfully designed. The schematic diagram of the circuit is shown in **Figure 13**. It is worth mentioning that the input sound frequency is doubled due to the thermoacoustic effect, and this needs to be compensated during actual testing, as described ahead. The drive circuit uses a USB port to apply power to the circuit for amplifying the AC signal and also to apply an up to 15 V DC bias to the graphene earphone. In this way, the graphene earphone can connect to a laptop for playing music.

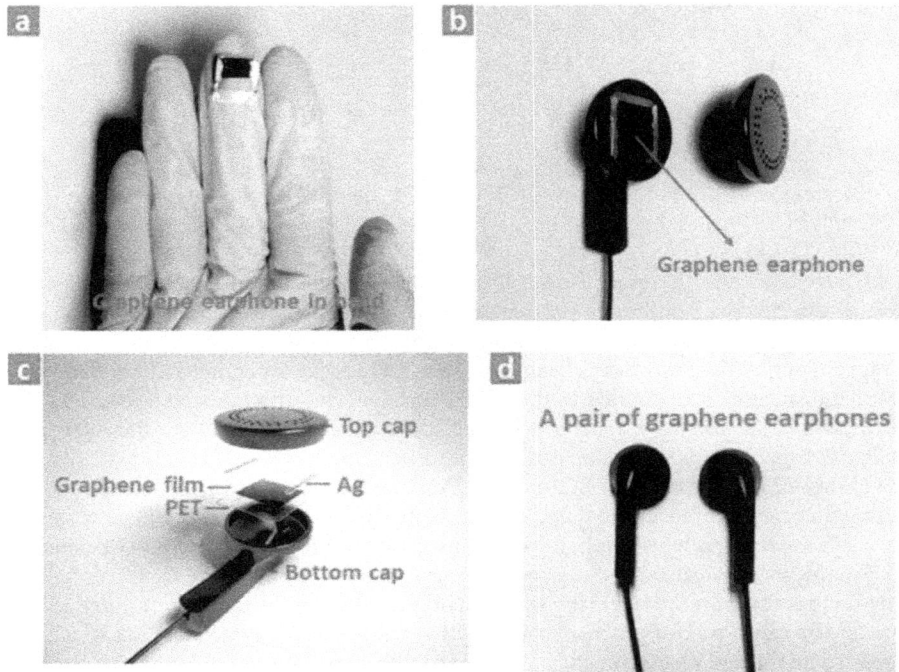

Figure 12.
The demonstration of graphene earphone [13]. (a) Graphene earphone in hand. (b) View of the graphene earphone in a commercial earphone casing. (c) Exploded view of a packaged graphene earphone. (d) A pair of graphene earphones in its final packaged form.

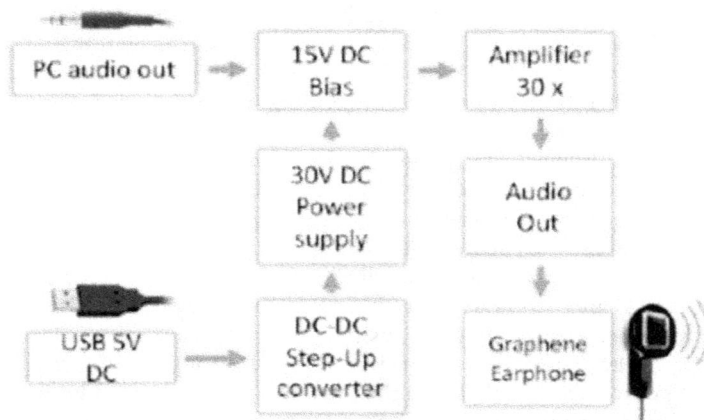

Figure 13.
Graphene earphone drive circuit structure diagram [9].

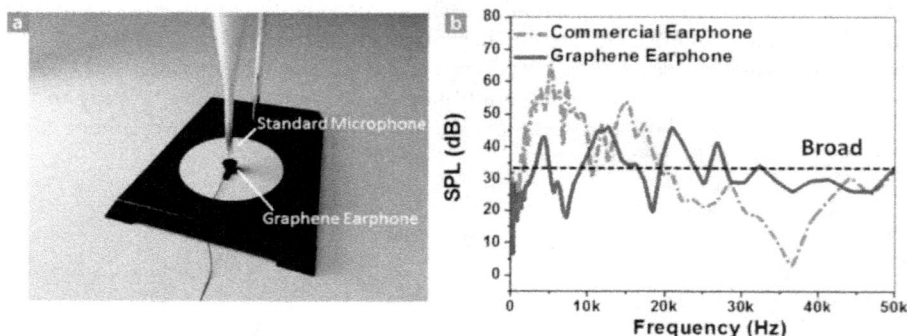

Figure 14.
Sound pressure and frequency characteristics of the graphene earphone [13]. (a) Experimental setup for the graphene earphone. (b) Sound pressure level (SPL) curves of a graphene earphone compared with a commercial earphone.

Figure 14a shows the graphene earphones acoustic test platform. **Figure 14b** shows the acoustic spectrum of the device. The graphene earphone is placed 1 cm away from the standard microphone. It can be seen that the graphene earphone has a flatter acoustic spectrum output than commercial earphone. Meanwhile, the sound intensity of the graphene earphone remains relatively stable. The graphene earphone has a fluctuation of 10 dB, which is much lower than the normal commercial earphone with the fluctuation of 30 dB. Especially, the spectrum can not only cover the 20 Hz to 20 kHz, but also cover the ultrasonic frequency band of 20–50 kHz. Animals are more sensitive to ultrasonic frequency than audible range, and graphene earphone has flat acoustic output in ultrasonic band. Therefore, graphene earphones can be used to train and control animal behaviors.

Finally, an application of the graphene earphone is demonstrated. The graphene earphone is fixed on the steel ring so that a dog can wear it, as shown in **Figure 15a** and **b**. The response test process is divided into the following sections. First, the dog needs to be trained so that it can recognize the "stand up" command when hearing the 35 kHz sound wave. The Chinese audio frequency of "stand up" was recorded first. Then, the audio was mixed with 35 kHz sine wave by computer software. The sampling frequency was 88.2 kHz, which was more than twice that of 35 kHz to ensure that the sampling was not distorted. The pure human voice was played to the

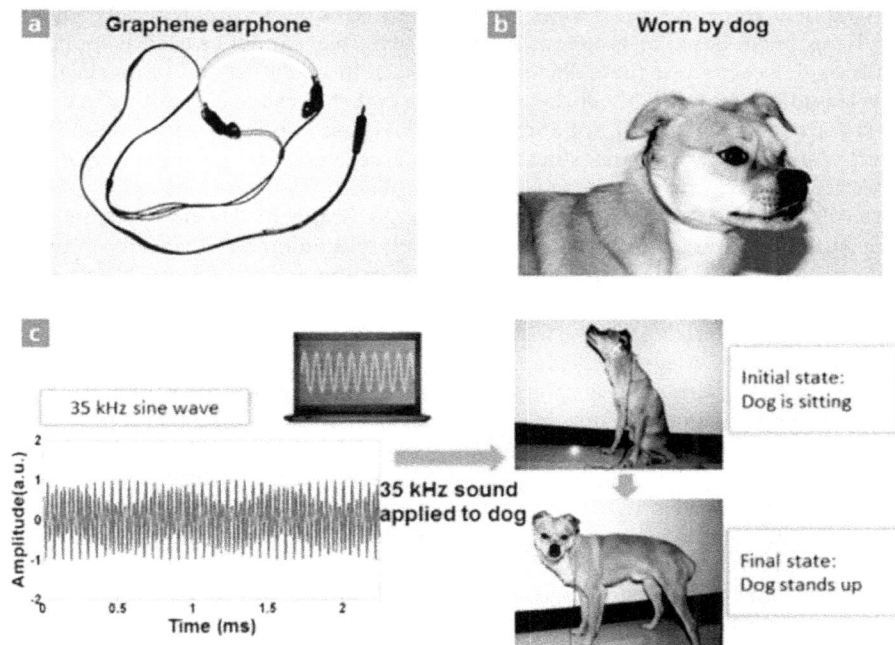

Figure 15.
Animal responding to ultrasound signals through graphene earphone [13]. (a) Graphene earphone for dogs.
(b) Subject wearing the graphene earphone. (c) Wearing a graphene earphone for the dog to establish a 35 kHz
sound wave and a "stand up" command to reflect the training process. The dog is initially sitting down. After
receiving a familiar 35 kHz signal, it stands up.

dog, and the dog will stand up, and then it will be mixed with 35 kHz sound waves. The dog will be given a food reward every time he stands up successfully. And then it went on to increase the proportion of 35 kHz sound waves, and to reduce the proportion of human voice. Until the final 35 kHz sound waves was 100%, the dog was able to stand up to prove that it was successful in establishing a 35 kHz sound wave to the dog and the "stand up" command. **Figure 15c** shows a demonstration of controlling dog behavior through graphene earphones. The initial state of the dog is sitting, and when the dog hears the 35 kHz sound from the graphene earphone, it will stand up, which indicates that the graphene earphone can control the dog's behavior.

5. An intelligent artificial throat with sound-sensing ability based on laser-induced graphene

The graphene sound source device based on thermoacoustic effect is introduced. Based on its piezoresistive effect, graphene can also be used for throat detection, which can be used as a good sound receiver. Therefore, based on thermoacoustic effect and piezoresistive effect, graphene is expected to realize transceiver integration.

At the present stage, devices with integrated sound and transceiver function are usually called ultrasonic transducers, which can only work in the ultrasonic frequency band or under water [14]. At the same time, these substrates are hard, not flexible and biocompatible, and not suitable for wearable applications. Therefore, studying the audible frequency band, the sound-flexible device with good flexibility and biocompatibility is of great significance.

Graphene force acoustic devices are mainly faced with two problems: (1) Sound receiving device based on graphene piezoresistive effect cannot be used as sound emitting device because these devices are wrapped in a polymer and the heat cannot be released into the air [15]. Or the device has a higher resistance, resulting in poor thermoacoustic effects [16]. (2) Sounding devices based on the thermoacoustic effect of graphene are also not suitable as sound receivers. Because these device are fabricated by using single layer or few layer graphene [17]. These kinds of graphenes are easily broken, which cause the devices damage. Graphene devices based on conventional processes and methods are difficult to simultaneously receive sound and emit sound using thermoacoustic effects and piezoresistive effects. Therefore, it is necessary to explore the realization of graphene transceiver sound integration based on new materials and new technology.

The method of laser scribed graphene (LSG) is described above. However, graphene reduced by this method cannot be simultaneously transmitted and received sound due to packaging reasons. In 2014, Lin et al. proposed a method for preparing porous structure graphene by laser reduction of polyamide material (PI) [18]. This method can be completed in a single step process without the need to drop coating GO, and the prepared graphene has a porous structure and a good piezoresistive effect. Because graphene materials have good thermal properties, they can make sound based on thermoacoustic effect. Therefore, the porous graphene based on laser reduction can be used to fabricate the flexible force acoustic device to realize sound transmission and reception integration.

This chapter introduces the process of preparing porous graphene by 450 nm wavelength laser. The porous graphene has the ability to integrate sound transmission and reception based on thermoacoustic effect and piezoresistive effect. Hence, a new intelligent artificial throat was prepared based on porous graphene. This device can be attached to the throat to sense the vibration mode of the throat of the deaf mute, and emit a preset sound when a specific throat vibration mode is detected. Therefore, the artificial throat can assist the deaf and mute to achieve sound, which will have potential application in biomedical, acoustic and other fields.

Laser direct writing technology promotes the fast growth of porous, and a low-cost and portable laser platform is chosen. **Figure 16a** is a schematic diagram of a laser processing platform. The 450 nm laser can directly reduce the yellow PI to the porous graphene. Then, the artificial throat based on the porous graphene has been integrated to achieve the functions of emitting and detecting sounds, as shown in **Figure 16b**. **Figure 16c** shows the working mechanism of the artificial throat. When AC voltage is applied to the device, the periodic joule heat will cause air to expand, thus producing sound waves. When a low bias voltage is applied to the device, the vibration of the throat leads to the change of the resistance of the device, resulting in current fluctuations. Therefore, this device can act as both sound sources and detectors. A coughs, buzzing or screams can cause throats to vibrate, which can be detected by LSG artificial throats, and then LSG artificial throats can produce controllable sounds. Therefore, LSG artificial throat can achieve the conversion from meaningless sound to controllable and pre-designed sound. **Figure 16d** shows the image of LSG at laser power ranging from 20 to 350 mW. It can be seen that there is no obvious LSG at the bottom when the power is 20 mW. The second black part and the fourth black part from the bottom, which are produced at the power of 125 and 290 mW, respectively, are chosen to show the SEM image, as shown in **Figure 16e–j**. It can be seen that regular ridge lines are formed from top to bottom along the laser scanning trajectory. The line width is approximately 100 μm, which is similar to the focused spot size of the laser. As the laser power increases, the morphological differences are significant. A polygonal porous carbon film appears at the power of 125 mW. However, when the power is 290 mW, more porous irregular structures can be produced. This is mainly

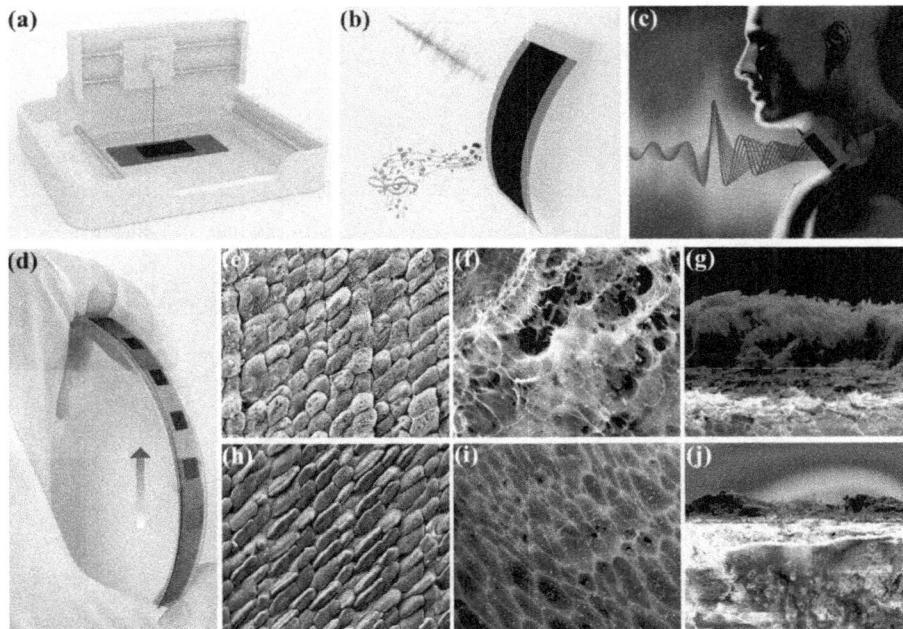

Figure 16.
(a) One-step fabrication process of LSG. (b) LSG has the ability of emitting and detecting sound in one device. (c) The artificial throat can detect the movement of throat and generate controllable sound, respectively. (d) Si_x LSG samples produced by 450 nm laser with different power ranging from 20 to 350 mW. (e) The morphology of LSG sample produced at 290 mW under scanning electron microscopy, scale bar, 150 μm. (f) The morphology of LSG sample produced at 290 mW under high magnification, scale bar, 5 μm. (g) Cross-sectional view of LSG sample produced at 290 mW, scale bar, 12.5 μm. (h) The morphology of LSG sample produced at 125 mW under scanning electron microscopy, Scale bar, 150 μm. (i) The morphology of LSG sample produced at 125 mW under high magnification, Scale bar, 5 μm. (j) Cross-sectional view of LSG sample produced at 125 mW, scale bar, 12.5 μm [19].

because the high power will lead to a sharp rise of the PI local temperature, thus breaking the C–O, C=O and N–C bonds, and causing the venting of some carbonaceous and nitric gases. Therefore, the porous structure of sample is formed due to the production and discharge of gases.

Four samples of LSG artificial throats are produced at different laser powers. The area of the LSG is around 1 × 2 cm^2. The laser powers are 125, 200, 290 and 350 mW, respectively. The average thickness of LSG is about 8, 22, 38 and 60 μm, corresponding to the laser power. **Figure 17a** is a test diagram of the device. The distance between the sample and standard microphone is 2.5 cm. **Figure 17b** shows the relationship of the LSG (produced under 125 mW) between input power and SP. The result indicates that the SP increases linearly with increasing input power, and a higher SP can be obtained at 20 kHz. **Figure 17c** shows the spectrum response of the LSG produced by different power. The input power is normalized to 1 W. It can be observed that the SPL gradually decreases with the increase of power. **Figure 17d** shows the comparison of theoretical curve and the experimental results. The experimental analysis matches well with the theory model. **Figure 17e** shows the experimental data and theoretical curve as a function of the thickness of the LSG under the frequency of 10 and 20 kHz. The SP is inversely proportional to the thickness of the sounding unit and the theoretical curve matches well with the experimental results. **Figure 17f** shows the stability of LSG. It can be seen that there are no signs of degradation or changes of SPL in the device performance in 3 hours.

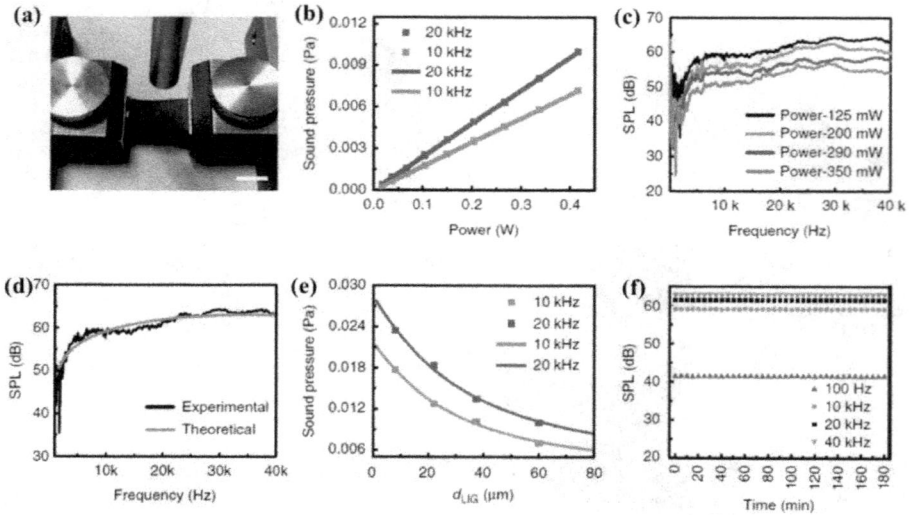

Figure 17.
(a) The LSG is clamped under a commercial microphone to test the performance of emitting sound, scale bar, 1 cm. (b) The plot of the SP versus the input power at 10 and 20 kHz. (c) The output SPL versus the frequency of LSG generated by the laser with different power. (d) The SPL versus the frequency showing that the model agrees well with experimental results. (e) The plot of the SP versus the thickness of LSG at 10 and 20 kHz. (f) The stability of output SPL over time [19].

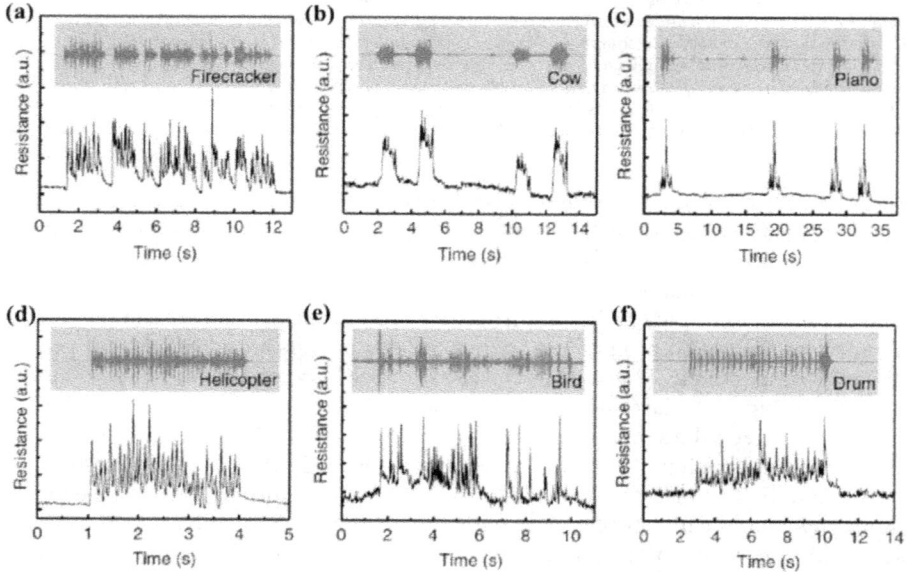

Figure 18.
Responses towards different audios from a loudspeaker [19]. The LSG is placed 3 cm away from the loudspeaker. The orange insets above indicate the sound wave profiles of the original audios. Relative resistance changes show almost synchronous response to profiles of the original audios when the loudspeaker plays the audio of (a) firecrackers, (b) a cow, (c) a piano, (d) a helicopter, (e) a bird and (f) a drum.

Except for emitting sound, the LSG artificial throat also has excellent responses when detecting sound. The 25 μm-thick PI can be chosen to produce the LSG due to obvious resistance change. The sample is fixed and the loudspeaker is placed 3 cm away from the artificial throat. The six kinds of audios including firecracker, cow, piano, helicopter, bird and drum are performed. **Figure 18** shows the resistance response of the device at six kinds of audios. Although the sampling frequency of this artificial throat is 100 Hz, which is far lower than the frequency of sound, it can be still noticed that the responses of the transducer are well synchronous to these original audio signals. Especially, the characteristic peaks are retained and reflected with high fidelity. Besides, the volume of loudspeaker has a great effect on the amplitude of the signal.

After identifying some kinds of audio clearly, LSG artificial throat is used to detect the vibration of throat cords. As shown in **Figure 19a**, the tester performs coughing, snoring and screaming twice in succession, and then the tester swallows and nods twice. After two consecutive tests, the test results are reproducible. In addition, swallowing and nodding can also cause muscle movement, which can also lead to changes in resistance. Fortunately, the waveforms of these muscle movements also have identifiable features. Different movement has its unique characteristic waveform as shown in **Figure 19a**, thus, the useful waveforms can be gotten by relying on the pattern recognition and machine learning. Interference from other activities can be identified and eliminated by multiple trainings in advance. Then, the tester makes the hums with four different tones as shown in **Figure 19b**,

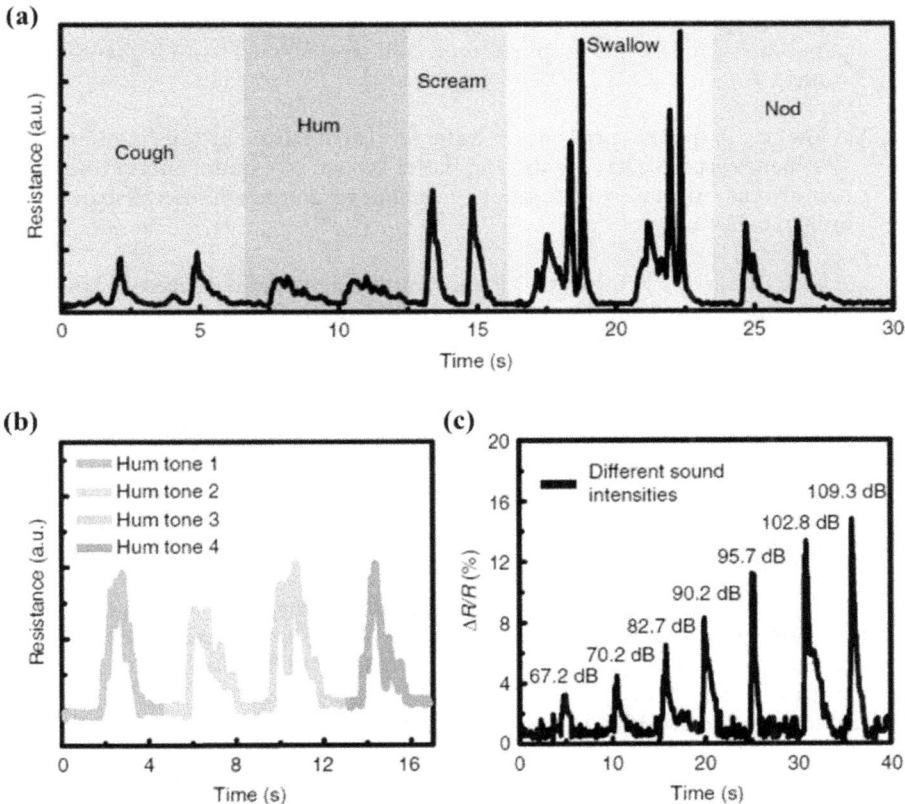

Figure 19.
Responses towards different kinds of throat vibrations. (a) The LSG's resistance changes towards the throat vibrations of the tester who makes two successive coughs, hums, screams, swallowing and nods. (b) The LSG's resistance change caused by four different kinds of hum tones and the hum tone 2 is same with the hum in (a). (c) The relative resistance change of LSG increases with the increase of the sound intensities of the hum [19].

it can be seen that different tones have different responses, increasing the diversity of dumb people's "language". Especially, the hum tone 2 is same with the hum in **Figure 19a**. Furthermore, as shown in **Figure 19c**, the resistance increases as the sound intensity increases, which is due to the increase in mechanical vibration of the throat.

6. Conclusion

Thermoacoustic device for the current low reliability, high performance and poor driving voltage problem, Ren's group successfully proposed and implemented graphene thermoacoustic devices with high performance, high reliability and low driving voltage. Besides, these devices have the advantages of low driving voltage, soft, transparent, thin thickness and wide band sound output, especially the extremely flat sound output in the ultrasonic frequency band. Exploring the application of graphene in the field of acoustics is the first time in the world and obtains the following three research results:

1. A multilayer graphene sound source device is proposed and realized. The sound output performance from 1 to 50 kHz is obtained. It is observed that there is a flat sound spectrum in the range of 20–50 kHz. The performance of multilayer graphene sound source device with different thickness is compared. It was found that graphene with thinner thickness had a higher SPL value.

2. A low cost graphene earphones at wafer level are realized by laser scribing. Graphene earphone has a wider and flatter acoustic spectrum output than commercial earphones. In addition, graphene earphone achieves control of animal behavior.

3. A wearable artificial throat that is manufactured in one step based on LSG has been implemented. The LSG device achieves the functional integration of emitting and detecting sound due to its excellent thermoacoustic and piezoresistive properties. The LSG artificial throat has a relatively broad frequency spectrum because of resonance-free oscillations of the sound sources. Besides, as a sound detector, the LSG artificial throat can capture the mechanical vibration of throat cords with a fine repetition.

Acknowledgements

This work was supported by the National Key R&D Program (2016YFA0200400), National Natural Science Foundation (61434001, 61574083, 61874065, 51861145202), and National Basic Research Program (2015CB352101) of China. He Tian thanks for the support from Young Elite Scientists Sponsorship Program by CAST (2018QNRC001). The authors are also thankful for the support of the Research Fund from Beijing Innovation Center for Future Chip, Beijing Natural Science Foundation (4184091), and Shenzhen Science and Technology Program (JCYJ20150831192224146).

Author details

He Tian[1,2*†], Guang-Yang Gou[1,2†], Fan Wu[1,2], Lu-Qi Tao[1,2], Yi Yang[1,2*]
and Tian-Ling Ren[1,2*]

1 Institute of Microelectronics, Tsinghua University, Beijing, China

2 Beijing National Research Center for Information Science and Technology
(BNRist), Tsinghua University, Beijing, China

*Address all correspondence to: tianhe88@tsinghua.edu.cn; yiyang@tsinghua.edu.cn
and rentl@tsinghua.edu.cn;

† These authors contributed equally to this work.

IntechOpen

References

[1] Novoselov KS, Geim AK, Morozov SV, et al. Electric field effect in atomically thin carbon films. Science. 2004;**306**:666-669

[2] Ho KI, Boutchich M, Su CY, et al. A self-aligned high-mobility graphene transistor: Decoupling the channel with fluorographene to reduce scattering. Advanced Materials. 2015;**27**:6519-6525

[3] Li D, Chen M, Zong Q, et al. Floating-gate manipulated graphene-black phosphorus heterojunction for nonvolatile ambipolar schottky junction memories, memory inverter circuits, and logic rectifiers. Nano Letters. 2017;**17**:6353-6359

[4] Sarker BK, Cazalas E, Chung TF, et al. Position-dependent and millimetre-range photodetection in phototransistors with micrometre-scale graphene on SiC. Nature Nanotechnology. 2017;**12**:668-674

[5] SMM Z, Holt M, Sadeghi MM, et al. 3D integrated monolayer graphene-Si CMOS RF gas sensor platform. npj 2D Materials and Applications. 2017;**1**:36

[6] Arnold HD, Crandall IB. The thermophone as a precision source of sound. Physical Review. 1917;**10**:22

[7] Shinoda H, Nakajima T, Ueno K, et al. Thermally induced ultrasonic emission from porous silicon. Nature. 1999;**400**:853-855

[8] Xiao L, Chen Z, Feng C, et al. Flexible, stretchable, transparent carbon nanotube thin film loudspeakers. Nano Letters. 2008;**8**:4539-4545

[9] Tian H. Graphene-Based Novel Micro/Nano Devices. Beijing: Tsinghua University; 2015

[10] Tian H, Ren TL, Xie D, et al. Graphene-on-paper sound source devices. ACS Nano. 2011;**5**:4878-4885

[11] Blackstock DT. Fundamentals of Physical Acoustics. Vol. 465. New York: John Wiley and Sons, Ltd; 2000. p. 440

[12] Vesterinen V, Niskanen AO, Hassel J, Helisto P. Fundamental efficiency of nanothermophones: Modeling and experiments. Nano Letters. 2010;**10**:5020-5024

[13] Tian H, Li C, Mohammad MA, et al. Graphene earphones: Entertainment for both humans and animals. ACS Nano. 2014;**8**:5883-5890

[14] Meeks SW, Timme RW. Rare earth iron magnetostrictive underwater sound transducer. The Journal of the Acoustical Society of America. 1977;**62**:1158-1164

[15] Park JJ, Hyun WJ, Mun SC, et al. Highly stretchable and wearable graphene strain sensors with controllable sensitivity for human motion monitoring. ACS Applied Materials & Interfaces. 2015;**7**:6317-6324

[16] Cheng Y, Wang R, Sun J, et al. A stretchable and highly sensitive graphene-based fiber for sensing tensile strain, bending, and torsion. Advanced Materials. 2015;**27**:7365-7371

[17] Tian H, Xie D, Yang Y, et al. Single-layer graphene sound-emitting devices: Experiments and modeling. Nanoscale. 2012;**4**:2272-2277

[18] Lin J, Peng Z, Liu Y, et al. Laser-induced porous graphene films from commercial polymers. Nature Communications. 2014;**5**:5714

[19] Tao LQ, Tian H, Liu Y, et al. An intelligent artificial throat with sound-sensing ability based on laser induced graphene. Nature Communications. 2017;**8**:14579